M000301863

Lessons from the Lobster

Lessons from the Lobster

Eve Marder's Work in Neuroscience

Charlotte Nassim

The MIT Press
Cambridge, Massachusetts
London, England

© 2018 Charlotte Nassim

All rights reserved. No part of this book may be reproduced in any form by any electronic or mechanical means (including photocopying, recording, or information storage and retrieval) without permission in writing from the publisher.

This book was set in ITC Stone Sans Std and ITC Stone Serif Std by Toppan Best-set Premedia Limited. Printed and bound in the United States of America.

Library of Congress Cataloging-in-Publication Data

Names: Nassim, Charlotte, author.
Title: Lessons from the lobster : Eve Marder's work in neuroscience / Charlotte Nassim.
Description: Cambridge, MA : The MIT Press, [2018] | Includes bibliographical references and index.
Identifiers: LCCN 2017038811 | ISBN 9780262037785 (hardcover : alk. paper)
Subjects: LCSH: Marder, Eve. | Lobsters–Nervous system. | Neurologists–Biography. | Neurosciences.
Classification: LCC QL444.M33 N37 2018 | DDC 595.3/84–dc23 LC record available at https://lccn.loc.gov/2017038811

10 9 8 7 6 5 4 3 2 1

To Eve Marder

It is a capital mistake to theorise before one has data. Insensibly one begins to twist facts to suit theories, instead of theories to suit facts.

—Arthur Conan Doyle (1859–1930), physician and creator of Sherlock Holmes

Contents

Foreword

Eve Marder

What was I thinking?

Some of you might already know that I am partial to essay titles with double entendre, nuance, or ambiguity. Here I answer two questions: Why did I agree when Charlotte proposed this book project to me? and What have I learned through this process about what I was thinking scientifically at various stages of my career? This second question isn't trivial, as we all know that human memory is reconstructive, and recollections, even those of a scientist, are likely to be incomplete at best, wrong at worst.

Those of you who know me personally might have been surprised to know that it didn't take Charlotte a long time to get me to agree to her project. This is despite the fact that I have complicated reactions to being the center of attention. I like it when people read what I have written. I very much like people to show up to talks I give. In contrast, in general, I find interviews difficult and sometimes painful. In thinking about the last few sentences I have written, it has struck me that when writing and speaking in my own voice, I have control, or at least semi-control, of my words. For better or worse, in those instances, I have framed the boundaries as well as the context of the communication. In interviews, profiles, and even more significantly in this book, it is someone else who has chosen her words to place me in her own context. I have often failed to read published interviews with me because I find it painful to do so. I have read all of Charlotte's book, but it has presented me with an enormous emotional challenge because it is about me, but in Charlotte's voice. I have occasionally made a scientific point, but I have tried my best to respect the fact that this book is Charlotte's, and her vision of what she wanted to achieve is hers. I have tried to assist her without interfering in her project too much.

So, why did I say yes? Charlotte won me over because she said she wanted to write a book that could be used to explain to teenage girls that one could

be herself and follow a career in science. She said she wanted to find someone who was well enough known to have a name but also someone who
had followed paths slightly less trodden. So Charlotte took advantage of
my educational feminism and success despite working on nonconsensus
problems and not being part of the most elite and privileged part of the
American scientific establishment. I said yes because I immediately liked
Charlotte and greatly enjoyed talking with her. Indeed, through this process, we have become friends, and I never take making a new friend lightly.

In addition to liking and respecting Charlotte, her process of working
on this book has been fascinating for me. We all remember selectively, and
I have always been interested in how well scientists remember their published work and why they did what they did. I never would have had the
patience to do what Charlotte did; she read all of my lab notebooks from
my early years. In the precomputer days, all data were preserved in bound,
paper notebooks, and I have kept all of mine and those of the entire lab
since 1970. The early ones are easier to understand because everything is
there. As time went on and the lab moved to digital storage devices, individual lab members were differentially good at making sure the paper notebooks contain summaries of the raw data. Although in the early days, I
wrote my own thoughts into those lab notebooks, as time went on, my
thoughts were less likely to be in our lab notebooks and more likely to be
on scribbles of paper that I gave to the other lab members who were doing
the work, thus lost unless they pasted those notes into their notebooks
(some did and some did not).

But what I have found so fascinating about this project is that, as Charlotte has tried to do an intellectual history of some of the scientific ideas
that have come from our work, she acted as a scientific historian and
showed me that the sequence of events in my memory was not always
correct. What I was thinking was not necessarily what I remember! She
found traces of ideas much earlier than I expected or recollected. She read
our published papers, not the papers in my head, and asked questions that
showed me how incomplete my recollection of many papers is, although
I had written them. Obviously, Charlotte was reading both our notebooks
and our papers, with her own set of questions and conceptual frameworks,
so her take on things is her take. Charlotte also interviewed and talked with
many friends, students, postdocs, and collaborators. It is fascinating what
those people remember, and how those of us who worked together have
memories that reveal different aspects of our shared past. I was surprised to
learn that my friend, Brian Mulloney, kept letters I wrote to him when I was
a struggling postdoc in the mid-1970s. Those letters, written in long-hand

in the days when people wrote letters and mailed them, remind me of the long correspondences characteristic of eras gone before.

Charlotte's book can also be read as an intellectual history of how science has changed since I started graduate school. My thesis papers were published as single author works, as was customary then. Today, the papers from my laboratory will usually have multiple authors, and many papers in the field involve many authors from many laboratories. Following the intellectual history of an idea or a finding in today's world is much more difficult because information travels virtually instantaneously. While in some ways this is good for the progress of the field, my story could not occur the same way in today's world. Those of you who are now age 12 or 16 or 20 and become scientists will have very different stories, but hopefully they will be filled with as much joy and satisfaction as mine has been. The thrill of seeing and understanding something, be it small or large, that no one has ever seen before is incomparable, and I wish it to all of you.

Introduction

Understanding the human brain has always been the Holy Grail of neuroscience. In recent years, with encouraging advances in research techniques, it has risen to the top of national research programs and is now very much in the public eye. Around the world, stupendous amounts of money have been committed to projects intended to describe the human brain or to simulate its activity on computers; one of the current big science, big money, neuroscience research projects aims to map the connections in the human brain and is to be funded with three billion dollars over the next few years.

The human brain rules mind and body by selectively recruiting networks from about ninety billion neurons and the connections between them are of overwhelming complexity. These neurons are exceedingly difficult to study because of their number and because of the obvious constraints on research on human subjects and human tissue. Therefore, the vital foundations for human brain research projects have been laid elsewhere in the analysis of much smaller networks with countable numbers of neurons. You have in your hands a book that explores forty years of research on just thirty neurons. Illuminated by her uncommon intellect, the study of a small knot of neurons found in lobsters and crabs has made Eve Marder a leading voice in neuroscience.

How is it possible that such a minuscule blob of cells has become a goldmine of information on the nervous systems of all animals? Intuitively it may seem unlikely, but all nervous systems contain similar neuronal circuits, although some of them use prodigious numbers of neurons. The "operating principles" are the same and many of the signaling molecules are the same, conserved through millions of years of evolution. Once an organism had developed a useful repetitive movement that depended on a certain conversation between neurons, or had successfully used a particular substance as a neurotransmitter, that innovation tended to be retained during evolution in the library of mechanisms that are now also found in larger nervous systems.

Over the decades, Marder has coaxed the little circuit of neurons into answering all sorts of questions, and most of the answers have turned out to be generally relevant to the neurobiology of all species, including ours. Her remarkable ability to extract general principles from detailed experimental results is at the heart of this story. This book focuses on Eve Marder's thinking about the organization of neurons in circuits and the complexity and flexibility they reveal.

* * *

Professor of Neuroscience Dr. Eve Marder has worked at Brandeis since 1978, when she joined the university as an assistant professor and founded her own laboratory. I first came across her much later, in London, on a warm summer day in 2006. I hadn't taken much notice of her work before because I was working in a mouse lab. That sounds like a weak excuse, and I'm not proud of it, but it's symptomatic of the unhelpful chasm between the worlds of research on vertebrate and invertebrate animals.

I was grappling with a new, exaggeratedly bureaucratic order form when a colleague breezed in. "Aren't you going to the one o'clock lecture? It's Eve Marder. She's right up your street." Escaping tedium but regretting my lunch, I grabbed my notebook. A woman dressed in lime green and straw yellow was already starting her talk as she stepped onto the podium of the Young Physiologists Symposium at University College London. She seemed to glow with intelligence and enthusiasm. In a couple of minutes, she had everyone engrossed in the research she was describing, and I thought I could hear jaws dropping as the force of her conclusions sank in.

Her research model was, and is, the awkwardly named stomatogastric ganglion, the control center for the crustacean stomach. It wasn't even in my vocabulary. But Marder's lecture dismantled some accepted certainties in the most convincing way. For example, everyone in that lecture theater knew the standard textbook descriptions of nerve cells and their basic characteristics, such as the electrical properties of their membranes. They were taken for granted as the normal baseline condition. But after years of examining her lab's experimental results, Marder had found that there was a range of such properties in individual neurons and that many of them could change to suit requirements. A neuron could "tune" itself to its target task, or return to work after some perturbation, without necessarily restoring properties such as membrane conductance to previous values. To her audience, these were novel and unsettling ideas.

She had also thought deeply about the lifetime of the lobster and its neurons, linking her research to practical consideration of what animals do to grow and survive in the wild. How do neurons, which are not renewed in an animal's lifetime, work together as they and the animal grow? How do neurons consistently fulfill their function while constantly renewing their components and responding to changing circumstances—a key question in studying the durability of memory? At the neuronal level, it seems, they don't all do it the same way, even within the same species. The two concepts of neuronal homeostasis and animal-to-animal variability were woven together in Marder's lecture. Sandwiches went untouched; the whole room hung on her every word and buzzed with discussions at the end.

I asked myself why no one else had observed and drawn together these phenomena, why more researchers weren't going back to first principles and questioning the received—and convenient—wisdoms in ways that changed the game. Of the many, many scientists making original discoveries, only a few are good at recognizing and explaining their significance. Marder seemed to me to go further; she had a particular insight, a flair for the big picture, and an uncommon feel for the realities of the natural environment of animals, something all too often overlooked in the research lab. Years later, I found I still wanted to know more about her, so I emailed her, proposing a meeting.

Marder accepted and, after a long discussion of my still woolly project, she agreed to be the subject of this "thought biography." It's a book that could only have been written in close partnership with its subject, and Marder has given me many long interviews, telling her story and exploring the thinking that led to each of her major contributions to neuroscience. She has allowed me the freedom of her lab and the use of her notebooks. I asked many of her colleagues and students for interviews, and they all responded with enthusiasm and evident admiration for her work. "Eve is just the most striking intellectual presence in the field," said one of them.

I describe this book as a "thought biography" for want of a ready-made term. It is not intended to be a biography. I have made no attempt to cover the events in Marder's life comprehensively, mention every colleague she has ever met, or list her guest lectures, prizes, or important positions in national institutions. Although her personal approach, her concern for her laboratory team, and her warm professional relationships are integral to the way she conducts her research and to the cohesion of the small scientific community that studies the stomatogastric ganglion, an examination of that aspect of her working life is outside the scope of this book. Marder is an effective champion, promoter, and supporter of women in science; that too has to be another story. Nor is my purpose to tell the story of her private life; her family is mostly absent from these pages. I have skated over her childhood, cherry picking the most telling memories that reveal something of her enquiring mind as a child and her intellectual progress at school because I think some of the influences that formed her intellectual style can be detected in them.

The core of my project is Eve Marder's work, which reveals the outstandingly original scientific mind and the distinctive approach, as intuitive as it is analytic, that define her thinking. A key characteristic of this approach is her mistrust of the hypothesis-driven research that has long been established as scientific method. It expects researchers to form a hypothesis and then design the experiments that might prove or disprove it. Marder prefers to "let the data speak, let the data drive the questions."

Marder's research results over the years are both detailed and broadly significant, but no one result has revealed a Higgs boson of neuroscience and no one result labels her. What she has done is marshal her masses of data into a big frame. Every scrap of information, expected, unexpected, or enigmatic, is a pointer for her, a shiver of the needle on nature's compass. Her discoveries are of fundamental truths in biology, statements of unsuspected basic principles that have been revealed as much by her creative thinking as by her laboratory's findings.

After her earliest research on the characteristics of the stomatogastric ganglion, its neurotransmitters and connectivity, Marder pioneered the studies that introduced neuromodulation. This term applies to the way the responsiveness of neurons can be altered by chemical substances reaching them, not through synaptic contacts with other neurons, but by other pathways. The neuronal circuits of the stomatogastric ganglion produce rhythmical patterns of activation, and Marder has shown that functional subcircuits of neurons can be formed or disbanded by the action of such neuromodulators.

In 1993, she initiated the development in her lab of the "dynamic clamp," a breakthrough in electrophysiological methods, now used in labs all over the world. It allows researchers to observe real biological neurons responding to inputs calculated by a computer using feedback from the neurons' responses.

Her lab has led the way in analyzing the characteristics of neuronal circuits, showing how they operate under a wide range of conditions affecting each of their neurons while keeping the networks robust and reliable throughout the lifetime of the animal. Marder recognized that individual animals of the same species may find different ways of running their circuits to provide similar output—multiple solutions as they came to be called.

Marder's lab is now recognized as a leader in computational neuroscience and is currently using both theoretical and experimental research to analyze the closely linked concepts of neuronal homeostasis, animal-to-animal variability, multiple solutions, and compensatory mechanisms. Marder has formulated coherent theories of these phenomena. The findings apply equally to the neuronal circuits of vertebrate animals, although for many years this was disputed.

I regret that I have had to be ruthlessly selective in telling Marder's science story; her long list of publications contains a wealth of fascinating and significant material that I wish I could have included. Alas, I am offering readers something like a "best-of" compilation disc. But then there is this question: Is the best yet to come? This book has been years in the making, and there is no obvious finishing line. Both Marder and I now recognize that not only has neuroscience made enormous progress during those years, but Marder has advanced in accomplishment and in status, her lab has thrived, her perspective on the past has evolved, and her goals for the future have changed.

* * *

Eve Marder is now in her late sixties, a friendly woman who naturally attracts affection and esteem and is renowned for her mentoring of younger scientists. She goes to the lab every morning, arriving around 10 o'clock. Her office door is always open, and researchers go in and out all day with questions and problems, new data and new ideas. Marder encourages this traffic, baiting her room with chocolate and biscuits to entice the team to bring her the most recent data from experiments. She says she loves data, gets "grumpy" if the results don't flow in, and can't understand people who don't want to scrutinize them immediately.

Two sides of Marder's office are lined with shelves holding orderly stacks of books and reprints of her published papers. Opposite them, her L-shaped working space is covered in a disorder of papers, journals, data printouts, and notes that transcends the word "untidy." A widescreen computer emerges from the chaos under a life-size cutout of Xena, Warrior Princess, scowling into the room. In contrast, her apartment on Boston Harbor, with its picture windows overlooking the water, is tidy and well organized.

I once asked her how she gets so much done and yet stays so relaxed. She said it's because she reads very quickly and can write a thousand words in a couple of hours. She added that she doesn't agonize over perfection because she gets bored easily. So how did she get through the years of repetitious experimental procedures, I asked. "Well, it's like craft, like needlepoint." Marder's apartment is adorned with gorgeous needlepoint cushions, and skeins of wool in mouth-watering colors are wound on a rack in a corner of her bedroom. She is famous for her love of color; her dazzling shoes are legendary, and there are a dozen pens with different colored inks on her desk. At any one time in the years I've known Marder, she has been juggling the demands of her lab with those of the university biology department, her teaching semester, the editorial boards of journals such as *Current Biology* and the *Proceedings of the National Academy of Sciences*, writing articles, myriad invitations to speak, and the time given to various committees and advisory panels. In 2008, Marder was President of the Society for Neuroscience, and she is also well known in the wider scientific community. She is a member of the National Academy of Sciences Council and a Deputy Editor of eLife.

No one was surprised when this influential scientist was one of the fifteen neuroscientists brought together to advise President Obama's B.R.A.I.N.* Initiative at its inception in 2013. She is now a member-at-large of the Multi-Council Working Group that oversees the long-term scientific vision of the initiative in the evolving neuroscience landscape. Curiously, however bold and forward-looking that vision may be, it actually obliges Marder to look backward. The B.R.A.I.N. Initiative aims to "map the circuits of the [human] brain, measure the fluctuating patterns of electrical and chemical activity flowing within those circuits, and understand how their interplay creates our unique cognitive and behavioral capabilities ..." If you delete the words "brain" and "cognitive" from this statement, it describes what research on the stomatogastric ganglion in lobsters and crabs has already achieved.

A circuit map or "wiring diagram" shows the basic anatomical framework within which each neuron is active. It shows what cells communicate with what others, but it doesn't show much about the signal passed between them—only whether it is excitatory or inhibitory, electrical or chemical. It's a case of knowing who is talking to whom but not hearing what they're saying. As Marder says, this sort of diagram "is absolutely necessary and completely insufficient." Her work, eavesdropping on those conversations, has revealed the organizing secrets of neuronal networks. Now the same information must be gleaned from groups of millions of human neurons, but starting this time from some basic, accepted rules and knowing that patterns similar to those found in simpler animals are likely to hold good.

*Brain Research through Advancing Innovative Neurotechnologies.

One question about Marder, about those forty years, that will probably occur to my readers is whether her work is the product of a particular time. This idea would in no way detract from her flair and brilliance, but was she perhaps absolutely suited to that era? Has the march of technology, which now promises to bring within reach the study of large numbers of neurons, overtaken her work? I think not. Marder has always embraced any discipline or new technique that can help to answer her questions and her ability to examine data while putting it into useful perspective and relating it to wider issues is exceptional. She views the arrival of "big data neuroscience" with an enthusiasm tempered only by her recognition of the need for sound biological intuition in exploiting it. She has certainly not run out of questions. Her work is in full swing, and many alumni of her lab continue to add to the story from their own labs. As one of her former postdoctoral researchers told me, "There's no doubt about it—we will all work on Eve's questions for the rest of our careers."

This book will have achieved its purpose if it shows that great science can still be done by the intelligence of someone like Eve Marder, without technical fireworks or lavish funding. Marder's data come from electrophysiological, anatomical, molecular, and computer techniques, using the same sort of equipment and methods as many in her field. She makes no big claims; she just puts forward her far-reaching thoughts, without fuss, without self-importance. Forty years ago, she started research on the neurotransmitters used by a small bundle of nerve cells. That research has led to work that now sheds light on the complex phenomenon of individual variability in human brains and the boundaries that separate robust healthy brains from those susceptible to disease and instability.

I leave the moral of this long tale to Marder: "If I had to think of any one thing I want your book to say, it would be that if you start working on almost anything, even nerve cells in the stomach of a lobster, and you do it thoughtfully and carefully, and keep going, you can end up with pathways into the most general and deep biological problems."

1 The Lone Reader

"How can I be so stupid?" is the first thought, the first moment of self-awareness that Eve Marder remembers; a very palpable moment, she calls it. She was three years old. The family was living in New York City, and her mother used to take her to a playground on 86th Street. It had vertical railings, and one day Eve looked at the bars and wondered whether her head would fit through. It did, but she couldn't get it back out.

She remembers crouching there, well and truly stuck, and thinking, "How stupid!" which is coolly self-critical for a toddler. She started crying, of course, and all the mothers started panicking, except hers. Dorothy Marder told them not to worry and, turning Eve's body sideways, pushed it the rest of the way out through the bars. Afterward she told everyone that obviously the head is the biggest part, and as far as she was concerned it was the only precious part. So it didn't matter if the body got scraped; it would heal.

I had asked Marder what she remembered as her earliest distinctly personal thoughts. I was hoping they would give me some insight into her thinking processes as an adult, as a scientist. And, like this cute story, I think they did. First, let me outline her childhood—remember, this is not a biography, so I can and will leave out any dates and events that don't advance my quest.

A year after the playground incident, and after her brother was born, the family moved to New Jersey, to Richfield, then a small town with a few shops and an elementary school. They had a garden apartment. The front door opened onto a dull stretch of lawn, but outside the back door the driveway was an invitation to everything important and exciting. When school started, Eve's mother took her to kindergarten a few times and then left her to walk the couple of blocks up the hill with the other children. Eve relished her new freedom to trot around on her own.

Soon Eve was given her reader's card for the town library, which was small and easy for a child to feel at home in. The books for young readers were grouped by grade levels. Eve read every book on the shelves, systematically, grade by grade, whatever the subject: children's biographies, famous explorers, science, history ... running far ahead of her age group. Only later, when she was surprised that school classes were teaching what she already knew, did she realize that some of the books she had so much enjoyed were textbooks for older children. By the time she was nine, she had finished the children's section and started on the rest.

The family often drove to Manhattan, and on those trips, sitting in the back of the car as it crossed the George Washington Bridge, Eve was struck by the hundreds of lighted apartments. There were people, whole families, behind those windows, all of them going about their business, each with personal thoughts and feelings. Eve was overwhelmed and disturbed by the countless numbers of human beings. In the myriad apartments, all the couples, like her parents, had found each other in the immense crowds. She had been told that you married that one special person, but—she had read about the millions living in Asia—what if your special person lived in China? How was life organized so that you could be sure not to miss your right person?

She told me she worried for some time and eventually asked her father. Eric Marder said, "Don't worry about getting married, just get a PhD first." And this surprising answer worked; Eve stopped worrying because she didn't have to deal with the problem yet. But she remembers the strong and uncomfortable feeling of perplexity, perhaps precisely because she failed to make sense of it at the time.

I asked Marder when she first realized she was clever. "Ah, Sputnik," she replied inscrutably. She was nine when the Soviet Union launched Sputnik on October 4, 1957, but she didn't see the now familiar images because the family hadn't got a TV set; her parents had a philosophical objection to television, and she grew up entirely without it. The Soviet Union's unexpected technological prowess caused paroxysms of consternation in the United States. It seemed to have caught America napping. A national brouhaha about science education reached all the way to her small elementary school in Richfield. After a new-style intensive science lesson, another child told Eve she was the smartest kid in the class and she was taken aback, having never thought about herself like that. School was always easy for her, but this was a weird new idea, and she wasn't sure what it meant.

Her mother now thought Eve would benefit from a new challenge. This was Hebrew School. The family was far from observant, but Dorothy Marder had been raised in a kosher household and Eve would later become aware of her mother's ambivalence about her Jewish identity. At the time, however, it was ostensibly to keep Eve busy and she accepted it quite happily.

The school was attached to the conservative congregation of the neighboring town, and the classes prepared children for the coming of age ceremony at which thirteen-year-olds become sons or daughters of the Law. Eve was two years younger than any of the other children. The first couple of lessons had her completely baffled. It was rote learning; they learned the names of Hebrew letters, and the sounds they represented, and copied them out. After learning several of them, letter and sound, letter and sound, "a light flashed on in my head and I understood absolutely what an alphabet was. Obviously I knew the English alphabet, but I understood then that as long as someone told me how to make the sounds associated with symbols, I could read anything. Very strong intellectual pleasure, my first. Because I knew I had understood the concept of an alphabet."

She had no further difficulty and enjoyed the classes because they used to divide the children into two groups and have competitions; it was quite entertaining, not challenging at all, and not at all what her mother had intended. They were not even taught what most of the words meant, certainly no modern Hebrew, just memorizing and chanting prayers phonetically. The fun came to an abrupt end one evening when Eve's grandfather, Eric Marder's father, was at the family dinner table. Born to an orthodox family in Poland, he had rejected rabbinical school and was resolutely nonobservant. But when the notion of a Bat Mitzvah for Eve came up, he was uncharacteristically angry. "Girls," he said, "don't have that ceremony; it's an American invention. You're either a Jew or not a Jew, and if you're a Jew, only boys get Bar Mitzvahed." Eve was stunned; she had never seen him interfere in his grandchildren's upbringing before, and in fact he never did again. Her father just said, "Okay," and that was that and the end of the classes. But the lesson was clear: you either do something and do it right, or you don't do it at all.

In any case, the family was moving. Eve's parents had built a house in Irvington, in Westchester County, New York. From there Eve would have had to go to a plain and earnest reform temple. "My mother had very good taste. She had taken me to a conservative synagogue where I used to go to Friday evening services, and I loved all the pomp and ritual. And I liked the cake downstairs afterward." Worse still, the reform temple's services were in English; even at eleven, Eve thought that made it all sound absurd. But she still feels that her mother harbored an undercurrent of resentment that her family's Jewish tradition was being abandoned, even though she herself didn't keep the Sabbath, didn't go to temple. Toward the end of her life, Dorothy Marder would talk wistfully and regretfully about the family having given up the Jewish tradition, but there was a temple right across the street from her house, and she didn't go to it. To Eve, this was characteristic of her mother's ability to accept two irreconcilable ideas at the same time. Dorothy Marder was extraordinarily erudite, and Eve appreciated the subtlety of her mind, but she could simultaneously and with no difficulty defend two positions that Eve found incompatible. "It's different," she would calmly say if challenged.

Eve's father had a degree in electrical engineering; this training and his personality both inclined him to rational thought. He thought deeply about any problem and derived his answer from first principles. When he wanted to learn to sail, he started by drawing out the forces on the sails, using vectors. When he learned Go, he studied with masters of the game in a New York club, but was beaten by his son at home. He asked the twelve-year-old, "How do you do it, you're not studying?" The boy just said, "I look at the board."

Her parents thus had strikingly different approaches to solving problems. Throughout Eve's childhood, she remembers incidents, sometimes leading to quarrels. "He could easily end up with an entirely wrong assessment of a situation by a totally reasonable process. So my mother was almost invariably right, and he was almost invariably wrong. He would get there by consistent logical thought, and she would get there in this very bizarre way, and that used to drive me crazy." Over the years, I suspect, Eve melded these two styles of thought and, although probably a naturally rational thinker, also heeds her intuitions with the result that she is not hobbled by logic.

* * *

After the move to Irvington, Eve started at her new school in the fifth grade. As a new entrant, she was given an IQ test. She whipped through it with ease because it was exactly the same test she'd done three months earlier at the end of the fourth grade. She had thought it was fun then, with vocabulary tests, shapes to manipulate, and number puzzles, so she remembered some of it and could now answer many more questions in the set time. At the end of the year, the whole class took an IQ test—and it was the same one again. "This time I must have knocked it into the stratosphere." The result was an interview with the school psychologist. Eve told her it was the third time she'd taken exactly the same test, but the psychologist didn't listen and gave her timed tests arranging blocks to make forms. Eve had the same blocks at home and had played with them all her life. So she made the shapes rapidly and told the psychologist she'd been doing it since she was two. Again, the psychologist took no notice and scored her highly. "She kept on saying, Oh my God I've never seen anyone do it this fast! That was my first criticism of adult understanding because I knew the assessment was completely flawed. I was trying to argue it wasn't measuring anything real because I'd played a lot with this toy." Next came a vocabulary test, and the psychologist got even more excited. "She was busy writing all this stuff down and deciding that I'm a genius. I knew I was smart enough, but I knew I wasn't a genius. I'd been given the same IQ test three times, and I read a lot of books. I kept on trying to tell her that she was fooling herself." Of course, it also gave Eve an insight into how little adults were listening to kids.

In the sixth grade, the school experimented with grouping by ability. Eve was placed with some other good students, and they were taught by an older teacher, close to retirement. Her method was to let the students work on the material at their own pace while she helped any children in difficulty. The classroom was always loud and raucous, and she didn't care. Eve liked her. "All she could do to the bright ones was not wreck us, whereas with the weaker students she could actually make a difference."

On the first day, the teacher gave out the sixth-grade math workbooks and told them to get on with it, check the answers, and when a section was clearly understood, go on to the next. Eve and two boys, racing each other and battling for accuracy, wiped up sixth-grade math in two months. They gave in their workbooks, and the teacher raised her eyebrows and said, "Oh." The next day she turned up with the eighth-grade text and workbooks, saying there was nothing in seventh-grade math. So the three of them worked on, and this time it was more of a challenge. They had to teach it all to themselves with no help from the teacher, and that slowed them down. They finished a couple of weeks before the end of the year, and the teacher gave them an algebra textbook. The school had a program for the quicker children to skip parts of the seventh-grade program but not the eighth grade. Eve went to see the teacher on the last day of class and said eighth-grade math hadn't thrilled her with its percentages and compound interest, and she didn't want to do it again. This time she was listened to. When her report card arrived, she had been promoted to the eighth grade and to ninth grade for algebra. So the twelve-year-old Eve skipped the seventh grade entirely, which was unusual at Eve's school, though not unheard of in the United States.

The eighth grade was a terrible year for her. Her two math mates had changed schools, so she arrived in the class on her own, the brainy one, promoted in advance of her years. The older girls reacted with cruelty, finding ways to deride and isolate her.

The algebra in the ninth grade program pushed her a bit, and she didn't do well on her first quiz. Then she worked really hard at it and scored 97 later in the term. She was excited and pleased because she had pulled herself back up, doing really well in work two years ahead of her age level. She told her father. He asked what happened to the other three points. That riled Eve and she decided never to talk about any of her grades at home. But she made the mistake of telling a couple of the kids in her class, and they took it as bragging. So Eve had to keep her satisfactions to herself and for the rest of the year was miserable, wanting to be included by the other students but always excluded, snubbed, and labeled "smart." In the end, she decided she was never going to talk to them either. But she remembers sometimes going home and crying.

That summer, Eve's parents sent her to a camp set in beautiful Vermont country. Camp Kokosing was liberal and interracial, expressing a counterculture that Eve had so far never come across. The experience was transformative. She met children from liberal families, from New York, Long Island, and Westchester. Some of them she remembers as quite unusual, and it came as a relief to find out that she was not the only one being shunned at school. There was folk singing, and she discovered Joan Baez and Bob Dylan. There was a camp newspaper. Best of all, Eve made friends and was suddenly happy. "It was after that summer that I understood there were other people around who were fine, and I didn't have to worry about my little local high school." At thirteen, she was just old enough to take the train to New York City, where she could see these new friends during the rest of the year. She went to Kokosing for four summers and it gave her what she missed at school. And it was here that she started thinking about politics and the civil rights movement.

The following years, in senior high school, were less harrowing because Eve had more perspective on her situation. She now had the confidence that she could make friends outside school, although in her class she was still isolated by the year's age difference and retreated, as before, into solitary studiousness. Her peace pact with the other girls was forged on the sports field. She loved sports, and she was solid and strong, with quick reflexes. In field hockey, the gym teacher put her in goal, which most of the girls were too scared to try. Eve recalls being tough, aggressive, and pretty good. Even as a ninth-grade freshman, she soon made the varsity team of older girls—there was little competition for goalkeeping. One day, walking back after hockey practice, she overheard another girl stop her boyfriend from teasing Eve, telling him to leave her alone, she's a good goalie. For the rest of her high school days, she wasn't part of their crowd, but they left her in peace.

That wasn't quite enough to assuage Eve's feelings and like many other gifted and excluded kids, she comforted herself by disdaining the system that snubbed her. The school had a club for sports accomplishments; points were awarded for passing tests on the rules for all the sports, for playing intramurals, and for the school teams. Students who racked up enough points were then admitted by a vote to the Leaders' Club. Its members refereed the intramural games and wore coveted green jackets. Eve qualified on the points but didn't get in, and she knew it was due to her lack of popularity. Eventually, at the beginning of her senior year, she was invited to join. "I refused, and I was the only person in the history of the school to get a varsity sweater, which you got for piling up lots of points, who was not part of the Leaders' club. My little private triumph."

Much more important were the inspiring teachers at the school. In this, she knows she was exceptionally lucky and that it was all due to the particular circumstance of her town. Irvington had been a mostly working-class town. Along the main street running to the railway station were small family-run shops, a gas station, a post office, a laundry. New houses were being built on the outskirts of town, occupied by middle-class families like the Marders, whose breadwinners commuted to New York by train. But Irvington was an old community that included a few large estates, and the wealthy owners were mostly absent. There were big stone houses with acres and acres of land. The owners paid sacks of property tax but didn't send their children to Irvington's schools, so the schools were inordinately rich for the number of children in the system. Thus, there were better than average teachers because the town could pay high salaries and because it was a comfortable place to teach with not too many students, just five hundred in grades 7 to 12. Eve got an excellent education there, probably matching that of any elite school.

She had a wonderful history teacher and some excellent English teachers, one of whom taught them grammar, real old school grammar, for which Eve is still grateful. A few of her school essays have survived the various family moves and bonfires. Reading them, I was struck by the way Eve's correct grammar and syntax supported formal thought processes, even when the thoughts themselves were immature. She had learned to gather up her evidence, lay out her arguments, and reach a formal conclusion.

An English teacher in the eleventh and twelfth grades had the students writing long well-thought-out papers. It must have been a surprise to receive less than five hundred words from Eve when an essay demanded two thousand and the subject could be anything a student liked. Instead, Eve wrote a complaint that, with all the other activities and assignments a student has, four days was quite insufficient to decide on a theme, plan the essay, gather illustrative details, consider structure and choice of words, and revise the text. The assignment would thus fail to give the student "any insight into the problems that face any sort of serious writer." She concluded, "It all adds up to an extraordinary lack of consideration on the part of the teacher giving such an assignment. It would be unfair even if one could profit from an assignment of this nature. But since work assigned is supposed to teach, and since nothing could be learned by doing the necessary rush job, the assignment becomes triply meaningless and triply unfair." Outgunned, her teacher replied, "Oh me, oh my. I'm sorry. I take it all back," and gave her 9 1/2 out of 10 for "good argument."

Eve remembers it all being a lot of fun, and they worked hard. She wrote a hundred pages for a term paper on Walt Whitman's poetry. For another paper she still remembers, she got every textbook she could find on American History and read writers such as W. E. B. Du Bois about what actually happened in the Civil War and about slavery. Her paper was a critical analysis of how those years were handled in the textbooks, comparing it to what the primary source writers had recounted. At sixteen, she had figured out that black history had been written out of the school texts. It was an invaluable training. Throughout her career, Eve's published papers have been exemplary; in fact she writes so well that I'm amazed I ever had the nerve to suggest writing about her. Until she writes her autobiography, I suppose I'm safe.

The first academic cloud over this pleasant success story came in the twelfth grade. She was in a class for the dozen best math students, and the teacher decided they should learn symbolic logic. It made no sense to her, whereas calculus had posed no problem. She talked to another student; he understood it completely. At that point, Eve knew she struggled with a certain type of abstract reasoning. She understood what a Boolean operation was, but the problems seemed outlandish. She finally decided that formal logic was completely illogical. She knew it had axiomatic rules and structures, and she could sort of play the game, but she had seen the writing on the wall. "It presaged what I later learned in college, which was that, as I went into more abstract quantitative work, I was going to hit a limit. I could certainly learn more math, but I was never going past a certain point. I was never going to deeply understand it. Important to know where you're going to hit limits, and I knew exactly."

On the other hand, Eve found that biology just fell into place in her mind; she loved reading about it and had an excellent memory for it. She had an outstandingly good biology teacher, Bernice Essenfeld, who recalls, "Eve was an excellent student with real insight into science. She was also a lovely person. When our school moved from one building to another, I came in on a Saturday to pack all the materials from the biology room and lab that had to be moved. It was a daunting task. I was surprised and thrilled when Eve unexpectedly appeared and helped me all that morning." Eve was her first student to take the national SAT biology exam early, in the tenth grade. Mrs. Essenfeld just gave her the advanced biology course textbook to read, and Eve got a perfect 800 score.

The twelfth-grade syllabus introduced key concepts of modern biology such as Mendelian inheritance, chromosomes and genes, the theory of evolution, cell structure and function, the respiration cycles, and photosynthesis. The pathways in these cycles, molecule to molecule, and the enzymes that help transform them to make a cell's energy are complicated and notoriously confusing, but at age sixteen, Eve had no trouble with understanding and retaining it all.

Then, as she was lying in bed one spring morning, looking up at the tree outside her window, the ineffable complexity of living processes overawed her. The tree was bright with young leaves, just stirring in the breeze, and it looked completely peaceful until the contradiction struck her. "There's this placid tree," she remembers thinking, "and inside every one of its cells there are all those electrons zooming around ferociously making all of its energy. How strange it was that hidden behind the exterior, all the molecules were working and the enzymes and all the little machineries, like factories."

She tried to imagine things at the scale of a single cell. Inside the cell, millions of molecules, about a hundred million in each cell, are moving ceaselessly in the crowd. There is no up or down inside a cell because at that size gravity has virtually no effect. The cell's membrane isn't what the word implies in our everyday life. It isn't a durable rubbery casing; it's made up of fatty molecules called lipids and has a consistency a little thicker than olive oil. But at the scale of cells and the molecules they contain, even water slows everything down as if it were some heavy-duty gel. Behind a lipid membrane, most molecules might as well be caged in by iron bars.

"It's stuff like that," she thought, "stuff of submicroscopic size that is keeping you going, day and night, rushing about its chemical business. Behind your everyday experience is a whole other world." She tried to get those two images fused in her head. She thought about it a lot, conscious that it was the deepest, hardest thinking she had ever attempted. After grappling for weeks with the contrast between the surface and the inside, she could finally reconcile that complexity—the phenomenal energy and ceaseless activity going on despite the image of calm. Nevertheless, she had found it disturbing: "It was just like looking at all those people behind the windows in New York." It was one of the two insights Eve had that year that were of lasting significance to her.

Mrs. Essenfeld taught her class about the nervous system in some detail, including nerve cell structure and function, reflexes, conditioned reflexes, and involuntary and voluntary behavior. Eve learned that nerve cells, neurons, are of different shapes compared with other cells, with long thin axons and dendrites projecting from the cell body. She was interested enough to read—and still remembers—an article about neurons published in *Scientific American*. I looked up the index for 1965 and found what she described. It was written by the eminent Sir John Eccles, who had shared a Nobel Prize two years earlier with Andrew Huxley and Alan Hodgkin for work on the synapse, the point of contact between neurons. Neurons transmit signals at synapses by releasing molecules of neurotransmitter. These molecules fit into much bigger assemblies of molecules in the membrane of the neuron receiving the signal on the other side of the synapse. The complex molecules in the membrane are called receptors. They change their state or shape on receiving a neurotransmitter molecule that they recognize. In various ways, this allows electrically charged ions to move across the membrane into or out of the cell. As a result, close to the receptors, the electrical properties of the membrane change. But the signal received at any one synapse consists of a few molecules of a neurotransmitter and is generally far too insignificant to activate the whole neuron; it is the summation of many active synapses that can excite the receiving neuron to "fire" and release a neurotransmitter in its turn. Eccles called this the "nerve impulse," but it quickly became known as the "action potential," a term that is less obviously descriptive, alas. Many older scientists harbor a real nostalgia for the writing style of those days. This is how Eccles ended his article: "This account of communication between nerve cells is necessarily over-simplified, yet it shows that some significant advances are being made at the level of individual components of the nervous system. ... We can be encouraged by these limited successes. But the task of understanding in a comprehensive way how the human brain operates staggers its own imagination."

These were new and exciting discoveries for scientists, let alone a high school student, and led to Eve's second tussle with contradictions. "I had a conversation with myself about what was up there in my head; it was just neurons. This conversation started in high school and went on for several years through college: how can sense of self exist when all you've got is neurons? Who is it that I am when all I've got is nerve cells?" Again she tried to face a profound contradiction, this time between the complexity of the brain's billions of neurons and the apparent unity of consciousness. It has to be said that in the following decades, many distinguished scientists and philosophers have devoted themselves to this problem, so far with no convincing explanation. "Finally I decided that was okay too. I spent a bit of time in those years coming to terms with it."

Today, familiar with the characteristics of Marder's work, it is perhaps too easy to see a theme running through these thoughts, a theme that I interpret as a need to analyze complexity, trying to understand how countless different entities can resolve themselves into a functioning whole. I asked her whether she was afraid of complexity. She answered thoughtfully, "Noo-oo. I'm not afraid of big problems and I'm not afraid of new approaches and I'm not afraid of complexity per se, but what I can't tolerate is a certain level of ambiguity, which is different in my mind. … I could imagine doing all sorts of things with lots of unknowns in them, but the idea of trying to understand how a large brain works makes me queasy just because I don't know what the components are. It's not that the problem is complex or difficult, it's not being able to find the problem."

That seems to sum up her difficulty with the placid tree or with billions of brain cells producing personality: she couldn't identify the question. Now, I think one of her major skills lies in identifying problems, putting her finger on the anomalous data that suddenly call for explanation, lining up a lot of question marks until she can frame the one lucid question they were all leading to.

Throughout her senior year, however, she felt her real intellectual engagement was with Shakespeare and Whitman—and civil rights. When she came back from summer camp, she talked incessantly about disarmament and civil rights and became active in the local movement. She thinks it was her enthusiasm that kindled her mother's commitment to similar causes. Dorothy Marder became a well-known photographer, chronicling the social activism of the late 1960s to the 1980s, such as the anti-nuclear and anti-Vietnam War movements. She is especially remembered for her photographs documenting the feminist movement and women's peace activism.

In autumn 1965, Eve went to college intending to become a civil rights lawyer. She didn't even consider science, although as a little kid when asked what she wanted to be when she grew up, she used to say "a scientist" because the first time she said it she got praise and attention. Eve wanted to go to a West Coast college, but her parents vetoed that idea. Instead, she went to Brandeis, then a young university, still small but with a growing reputation. It had been founded in 1948 at a time when Harvard and Yale ran quotas to limit the number of Jewish students and faculty.

In that heyday of the civil rights movement, Eve signed up for politics. She also took French, psychology, and humanities; she took none of the sciences, but mathematics was inescapable. Here, unfortunately, she hit the limits she had seen on the horizon in the twelfth grade. "They had three levels: Math 12 for people who were going to be math majors, Math 11 for science students, and Math 10, which was calculus for the great unwashed—at that time, every student had to have calculus. So I took Math 11, but that year the math department decided to teach formal analysis. It was very abstract, and I realized later it had been completely inappropriate for science students at our stage. I worked very, very hard. I struggled. I got a B, but I knew I wasn't really understanding it. We never saw a $\partial x/\partial t$ in the whole year, just functions with formal proofs. So the only calculus I knew at the end of the year was what I had learned at high school. That was probably one of the bigger tragedies of my scientific career: it just stopped me going on in math. Many years later, I had to teach myself the basics of differential equations." Out of necessity, she has since overcome her loss of confidence. Today the Marder lab routinely includes mathematicians and is known for computational as well as experimental neuroscience.

Eve loved the politics course, and at the end of her freshman year she declared as a politics major. Now she would not have to take any more science courses, ever. But somehow she felt she would miss biology. Eve is always a practical thinker, and she decided to keep her options open. She thought it would be easier to go to law school with a biology degree than to go to a biology graduate school with a degree in politics. Accordingly, in her sophomore year she took biology as well as politics. And she took the honors chemistry course for students majoring in science because she had placed out of the standard course by taking an exam when she entered Brandeis. On top of that she took a course on the poetry of Joyce and Yeats and a philosophy course. That made five courses, two of them with laboratory work. She had a ghastly schedule and the courses didn't inspire her much. Even the biology course was a disappointment, covering little that was new to her.

Her second politics course, however, changed her life. Eve told me about it with a shudder. It was taught by a froggy-looking man with an irrepressible love of detail. A heavy black textbook with double-column pages described all the post-World War II political parties in every country in Europe, with lists of their names and acronyms and summaries of their platforms, year by year. Such and such socialist party wasn't really socialist, these progressives were really right wing, and somebody else's newly founded reform party was really the Left, and the other reform party was actually on the Right. She hated it. And then there was the final exam. Eve had had mononucleosis and missed some classes. She was tired and worryingly behind on her reading. Two days before the exam, she sat down with the black book and read it from beginning to end. Then she remembers walking from the library to the exam hall trying to keep her head still because she felt as though the facts might spill out. "As I handed in my paper, I pushed the delete button in my head—interesting in terms of memory because I literally obliterated the whole course, it was only there in temporary storage."

A week later, she switched her major from politics to biology. Now she had to catch up on organic chemistry, an essential part of the biologist's toolkit. The only possibility was a Harvard summer school class that crammed a year's course into eight weeks, including the laboratory work. Although Eve was recovering from mononucleosis and still hadn't got her energy back, she had to take it. The course was in an old building with no air conditioning. There were big old-fashioned steam baths in the laboratory, so the temperature was usually around 95°F. A strenuous routine of morning classes, afternoons in the laboratory, and weekly tests exhausted her.

There was also unwelcome and, for Eve, unusual pressure. To transfer a course credit from the summer school, which would qualify her to major in biology, Brandeis required a minimum grade B. The course was full of very able students, mostly from Harvard and MIT, who were picking up organic chemistry to satisfy medical school entry requirements. The ability of these students was directly relevant to Eve's stress: all the students' grades for the weekly tests were plotted on a curve, and the mean had been arbitrarily assigned a C+. Bs were given for a certain number of points above this mean. Eve fought the curve, and for the first three or four weeks she managed an A or A–. But then, the day before the next test, her grandfather died. Her father phoned and told her to take the exam and come home afterward. She missed a few days of the course and never really caught up, struggling in a blur of heat and steam. 1967 was a solitary summer for her; she didn't have time to see friends.

In the end, she managed a B, and that was adequate because the grade wouldn't be transferred to her record at Brandeis. She washed her summer dresses and packed up to join her parents on holiday in Provincetown. When she unpacked her suitcase, her cotton dresses disintegrated. Horrified, Eve thought about the solvents she had been using in the lab. If they did that to her clothes, what had they done to her?

In autumn 1967, beginning her junior year, she shared an apartment with a close friend, a literature major. They talked about the friend's little sister who was being called autistic but at six was severely handicapped and getting rapidly worse. As well as her clever brother, Eve had a young sister, eight years old and thriving. Eve was dismayed for her friend. At the time, nobody knew what caused the condition or how to treat it. With hindsight, Marder thinks it might have been Canavan's syndrome, a hereditary neuro-degenerative disorder. The roommate decided to take a course in abnormal psychology. She came back from the first class to tell Eve she absolutely had to join it—on the grounds that the professor was *so* cool and well dressed and had a dueling scar.

Eve went to the second class and an elegant professor entered, wearing a three-piece suit, and sure enough he had a scar on the side of his face that looked just like a nineteen-year-old's idea of a sabre wound gallantly acquired. He was a good lecturer too, entertaining and original. Eve signed up for the course.

The romantic notions turned out to be mistaken in weaponry though not in gallantry. The scar was the result of injuries in a naval minesweeping mission in World War II. Professor Brendan Maher was a distinguished pioneer of psychology who had introduced what scientists like to call "a paradigm shift" in the discipline, bringing experimental enquiry to what had been a descriptive science.

In a class about schizophrenia, Maher explained that the prevailing "double bind" hypothesis, of childhood emotions caught up in conflicting cues from the mother and leading to an escape into delusion, did not seem to be supported by evidence such as twin studies. He told the class it was possible that schizophrenia was a biological condition; the inability to deal with sensory inputs, for example, might be caused by deficient inhibition in the brain. He said that this was not yet generally accepted and perhaps even something of a heresy. Eve was suddenly passionately interested.

When the term paper came up, Eve says, "I decided as a biologist I should write a paper on that theory. I go off to the science library, and I start looking in the stacks for anything on inhibition in the brain. There's not much known at this point, it's 1967, you know, and inhibitory neurotransmission isn't described fully until the late 1960s. But it was pretty clear there was inhibition in the brain, and in order to make sense of it, I started reading neuroscience." She probably read most of what was known at the time about inhibition. As she tells this story, we shake our heads: no student could do that today. She got an A for her paper.

"That's how I ended up a neuroscientist! No, I don't *really* think that the double-columned tome drove me out of politics or that the random comment that Brendan Maher made about inhibition in the brain made me a neuroscientist. But sometimes there are these accidental moments that just hit right at those interesting points in time." During those long sessions in the library, Eve decided she wanted to study the nervous system. Neuroscience was still a relatively unfrequented backwater within biology departments; the word "neuroscience" had first been used in 1964. The next semester, she took the only relevant course on offer—actually only half a course on neuroscience; the other half concerned muscle structure and function. The course was mostly populated by seniors: about thirty men, one other woman, and Eve, who was a junior. It was daunting, hard work, and because she felt intimidated in this class, she didn't say a word. After the midterm exam, the professor gave her an A grade and showed surprise that this silent student had been taking in anything at all. In fact, Eve had been drinking it all in, loved it, and was particularly fascinated by the work on neuronal signaling and structural biology.

In an enjoyable senior year, Eve took a dance class, a film class, courses in developmental biology and some poetry, along with her senior research class in biology. There was of course no specialized neuroscience laboratory, so she went to a biophysicist, Andrew Szent-Gyorgyi, who was working on the structure of the molecules in muscle that make it contract. "He had nothing to do with neuroscience, but he let me do an honors thesis in his lab. Doing anything in a lab was better than nothing because I'd never been in a research lab. Those guys taught me what a magnetic stirrer was, everything like that." Szent-Gyorgyi's wife, who worked in the lab with him, was kindly forgiving of Eve's lack of experience and trained her in the basic techniques. She was the only model of a woman scientist Eve knew— except for one other. At this point in our conversation, Marder looked off into the distance and declined to name names. "She was an extremely tall, frightening, abrupt witch of a woman, really off-putting. You didn't want to be like that." And so the picture of the sweet clever woman working in her husband's lab became the goal. You met some really cool guy and you married and then helped with his research. Fortunately, this self-effacing dream didn't last long.

In the academic year 1968 to 1969, students all over the country were protesting against the Vietnam War. The Brandeis campus boiled over with sit-ins and strikes. Eve couldn't possibly be a scab. Her way out was to work in the lab in the evenings and on weekends, having convinced herself that these hours didn't count: if she worked in her own time, she wasn't breaking the strike. Szent-Gyorgyi had a different and much more serious attitude toward politics and political engagement, shaped in Hungary. He laughed, or perhaps scoffed, at the ruse, but he strongly advised her not to go to Berkeley for graduate school because of the danger of being caught up in politics.

Her Brandeis professors knew there weren't many options in neuroscience because few universities had any neuroscientists at all. They advised her to become "a regular molecular biologist" first. Brandeis was resolutely modern and biology was molecular; no one was interested in all those plants and animals. Eve was determined to get into a West Coast graduate school because this time she wanted to be far from her family. She did make one exception; she applied to Harvard because of its renowned Neurobiology Department. She got an interview but was turned down. Stanford, she soon found out, was a lost cause because of its quotas: only two women in a class of twelve, and only a single entrant from any one undergraduate college. The Brandeis slot had been quickly taken by one of her friends. She turned down the offer of a place in Physiology at Berkeley, not because of the danger of politicization but because they told her she was their strongest applicant. That didn't promise an inspiring level; Eve knew she was smart, but she also expected plenty of students to be smarter.

The University of Oregon also accepted her, and so did the University of California at San Diego. The San Diego campus was a new addition to the university. "It sounded exciting," Eve thought. "After Brandeis, it felt right to be rejected by the structured elite, and to go to this new institution where you could go your own way, it all depended on you, nothing was going to be given to you just because you were part of an established tradition."

So San Diego it would be. Then Marder laughs and admits that the well-known, winter-long rain of Oregon lost out to the lure of life in California. "I had this fantasy. I was going to do an experiment, put something in a centrifuge, go off to the beach, come back, take it out of the centrifuge. ... This really nice idea of what it would be like. And in some ways, it was very much like that!"

2   First Findings

A flood of women swept into life science graduate schools in September 1969. The draft exemption for male PhD students was ending, and far fewer men could take shelter from the Vietnam War in graduate studies. The universities were taken by surprise and had to scramble to fill places. Suddenly almost half the intake in the life sciences was female. In 1968, the Biology Department at San Diego had taken two women in a class of thirty. The next year there were thirteen. "There was a giant uproar!" Marder chuckles. "We, the women, arrived, and they were completely hysterical; they thought civilization as they knew it was over. By the end of that year, everyone had forgotten. There were good students who were women, good students who were men. The next year they just accepted fifty-fifty. In two years, entry into virtually every biology program around the country was gender neutral."

When we started to talk about the early years of her career, Marder singled out this watershed moment as being far more important for American women scientists than anything feminism could achieve. Because of their numbers, she says, her generation, unlike earlier women scientists, has had less need to be exceptionally tough, overdedicated, and often justifiably sour. And no need to become the unsung heroine of a husband's lab.

In 1969, the San Diego campus was nine years old. The Biology Department didn't offer graduates many neuroscience options, but there was already a neuroscience PhD program in the medical school just opposite the Biology Department; the University of California San Diego was on its way to becoming the neuroscience powerhouse that it is today. Although Marder was friendly with medical students and neuroscience PhD students—the campus was still too small for them not to know each other—she stuck to her department's program. In any case, she had no idea what research she wanted to do, and the advice of her Brandeis professors to get a PhD in molecular biology before specializing still rang in her ears.

All new graduate students in the Biology Department faced a year in "rotations," a series of short assignments in different laboratories intended to show them some of the current research fields and techniques before they settled on their own projects. While Marder was spending her autumn weeks in a plant protein lab, she heard the welcome news that the department had just hired its first neuroscientist. She hastened to see him in his temporary office and got his casual assent—he was a young man of few words—to a rotation in his lab as soon as it was set up. After a second rotation in protein biochemistry, toiling on what she calls "real hard-core chemistry," she went to Allen Selverston's new lab.

Before his post at San Diego, Selverston had been working on a crayfish model. Crayfish (and lobsters too) have a consistent, "stereotypical" way of escaping from danger. Rhythmic contractions of abdominal muscles snap the tail under the body, and the crayfish rapidly propels itself through the water—backward. Selverston was trying to describe the neurons that trigger these movements. He had published a paper that Marder thought beautiful, showing neurons made visible under the microscope by filling them with a yellow dye. Now, to clear up discrepancies between his findings and those of another scientist, he was working on the American lobster, *Homarus americanus*. He kept his lobsters in a bathtub full of seawater in a cold room that also housed algae. A green flashlight had to be used to handle the lobsters so that the algal photosynthesis regime wouldn't be disturbed. Marder remembers lugging ten-gallon carboys of seawater up the hill to the campus from the Scripps Institution on the beach. It was not a practical solution. For six weeks, she worked in Selverston's lab trying to stimulate axons in the lobster nerve cord. Like many rotation projects, it never got results and ran out of time.

Her next stint was in a tissue culture lab. Here, Marder was exasperated by the male chauvinism of her supervisor, who kept asking why she didn't stay at home and have babies. "After the twentieth or so lecture of this type, I told him I was spending the rest of my rotation in the library because it was clear he didn't want me in his laboratory." Her exasperation extended to the whole system. She decided four rotations were enough and abandoned any lingering idea of a PhD in molecular biology. Why would she spend years in some other biology specialty to prepare for neuroscience? "Screw that," she thought, "I'm going to do it now."

And so, in the summer, she took a neuroscience course at the Catalina Marine Biology Institute. When she came back to San Diego, she moved into Allen Selverston's lab—in his absence. Selverston had also gone off for that summer of 1970, to the Bermuda Biological Station, to work on a new preparation in a different kind of lobster. Marder had never formally asked him to supervise her work so he must have been somewhat surprised on his return to find her ensconced in his lab and engrossed in her reading. Selverston's attention, however, was focused on his change of preparation and of model animal. The tail-flip escape mechanism was replaced by a group of neurons called the stomatogastric ganglion and spiny lobsters definitively ousted both the American clawed lobsters and the crayfish. The bathtub was abandoned. Selverston made arrangements with a fisherman to supply lobsters and with the Scripps Institution of Oceanography for the use of their holding tanks.

For biologists to change the species they work on, the "model" for their research, is not a trivial matter, but Selverston was young and he had picked a winner. Neuroscience was also young, and the dominance we see today of models such as mice, fruit flies, nematode worms, or zebrafish was undreamed of. In fact the hunt was on all over the world for suitable models to study. Most animals have some easily observable sensorimotor reactions: the animal can be touched or startled in some way that always makes it flinch or try to escape. If an animal has a simple network of neurons that reliably activates this stereotypical behavior, then it is possible to separate out the cause from the mechanism. Researchers can identify the sensory nerve that detects the "stimulus" and the neural interchange where the message is passed on to a motor nerve that activates muscles. Desirable animals would have to be easily and cheaply available, and their behavior of interest would have to be controlled by a group of neurons accessible to laboratory procedures. And preferably those neurons would be large enough to manipulate under a microscope, allowing the investigator to stick fine glass electrodes into them to monitor their electrical activity. It made a long wish-list that pointed to the invertebrate world. The archetype of large amenable neurons, with an easily visible axon called "the giant axon," was the one that triggers the squid's escape response. That was the model for the trailblazing, Nobel prize-winning work by Hodgkin and Huxley in the 1940s. So candidate animals included sea hares, sea slugs, and various crustaceans such as lobsters.

In the late 1960s, Don Maynard had astutely recognized the crustacean stomatogastric ganglion as particularly promising because it was an independent neuronal network with a countable number of neurons, an apparently simple single input nerve, and an output to accessible stomach muscles. Maynard introduced this preparation to other neuroscientists, including the three pioneers, Allen Selverston, Dan Hartline, and Maurice Moulins, who were to become what you might call the founding fathers of stomatogastric ganglion studies. Maynard didn't live to see the stomatogastric ganglion adopted in numerous labs, supporting influential studies and major contributions to neuroscience. He died in 1973.

Few of us have ever given much thought to lobsters and crabs other than as attractions on a menu. Even fewer of us have wondered how they eat their own food. Lobsters break up their food with teeth—not in the mouth, but in the stomach. The "teeth" are cartilaginous ridges set at angles to each other, making a "gastric mill." The lobster's rocking, rolling, grinding, and filtering digestive process calls on sets of muscles to act together in coordinated movements. Those muscles are controlled by a group of neurons, the stomatogastric ganglion, which lies on top of the stomach, an inviting package ready for investigation. The neurons in the ganglion also signal to each other and synchronize the rhythmic churning. Crucially for this story, there are only thirty of them (figure 2.1).

This little system had the further advantage of usually staying "alive" in saline for hours (and nowadays longer, sometimes even for weeks) after careful dissection and preparation. So here was an in vitro preparation of a complete network that could be pinned down in a dish and then go on producing fictive motor neuron patterns that looked a lot like the in vivo ones. It was an especially happy choice for West Coast labs because the local lobster turned out to be easy to work on. *Panulirus interruptus* is a spiny lobster that lives in warm coastal waters like those of the Pacific at San Diego,

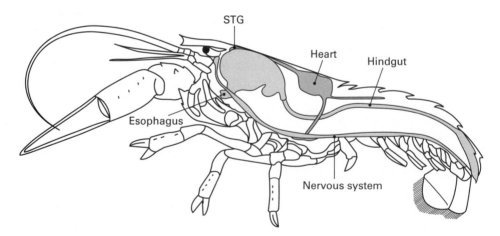

Figure 2.1
Shown here in the Maine lobster, *Homarus americanus*, the stomatogastric ganglion (STG) lies in the ophthalmic artery that runs from the heart to the eyes over the top of the stomach.

and its stomatogastric ganglion has a good, vigorous, spontaneous activity in a dish.

I can't overemphasize the sheer luck, for Marder, of Selverston's choice. Graduate students are inevitably influenced by their supervisors in their own choice of model. Marder joined Selverston's lab because he was doing good and interesting work when she met him, but also, and principally, because it was the first neuroscience lab she had known. She says she knew she could have joined other labs, studying cats, for example, but just didn't want to do that kind of work. She had already recognized the potential rewards of delving into a small but complete system with a countable number of cells. The stomatogastric ganglion model has been ideal for the detail and depth of her research. I am not suggesting that she would not have made an important contribution to biology under other circumstances, but this was a marriage of scientist and model made in some sort of laboratory heaven.

In Bermuda, Selverston had learned Don Maynard's dissection, which kept the lobster's stomatogastric ganglion intact while cutting it cleanly out of the body and then away from the stomach. The technique had the reputation of being incredibly difficult. Back in San Diego, Selverston took on a postdoctoral researcher, Brian Mulloney. The two men would work at the dissection for eight or nine hours and then start experiments on the ganglion at 7 or 8 o'clock at night, going on until the early hours of the morning. It was seen as a macho feat, and Marder avoided these sessions at first. She spent her days in the lab getting to grips with the techniques of electrophysiology, using the even smaller cardiac ganglion.

I interviewed Marder at length about the next three years, trying to reconstruct how she had picked out her research subject from the thicket of unanswered questions she faced and, of course, how she dealt with the challenges of experimental biology. Sometimes she would shrug and say she couldn't exactly remember. Fortunately, she has kept all her lab books. The top shelves of her lab at Brandeis hold them, in order, with all those used by others in her lab, up to number 896 as I write. At the far left, a fat notebook with a mottled brown cover and black taped spine is kept in a Ziplock plastic bag. It's labeled *#1—February 1971 to September 1971 Habituation*, and the squared pages are yellowing at the edges.

I started, naturally, on page 1: "In an effort to familiarize myself with the techniques and problems, the first thing I want to do is repeat Brunner and Kennedy's experiments on habituation in the motor giant [neuron] of the crayfish." Habituation to a repeated stimulus results in a diminished response, whereas a heightened response to a pattern of stimulation is called facilitation. Many neuroscientists were interested in habituation and facilitation because these altered responses hinted at the fascinating possibility of some sort of learning at the level of single neurons. Brunner and Kennedy had found that the axon of the crayfish giant motor neuron habituated at stimulus frequencies of ten per minute or slower and facilitated at the much faster frequency of one per second. Looking down the page, I found that Marder was thinking beyond the technical challenge of replicating the work. Yes, it had been shown that the axon habituated, but why? "I would like to ascertain whether the synapse is actually responding to the time since last stimulus or whether it is responding to a mean frequency."

I turned to page 2 and was astounded—and enchanted—to find that she was already trying to make links with the crayfish's natural behavior. As I consider this concern with real function, this surefooted leap from neuron to living beast, a hallmark of her work, it was remarkable to see that it was explicitly a part of her enquiries so early. In the midst of all the intimidating newness of laboratory techniques, faced with the immediate imperative of becoming competent, nevertheless she wrote, "Speculation concerning how motor giant habituation might function in terms of the animal's behavior. ... The motor giant is activated by the lateral giant command fiber ... I think it is activated as a fast escape response. As long as the lateral is firing fast (animal needs to get away fast—with little fine control of movement) the motor giant will fire (facilitation in fact). However, as soon as the laterals slow down, the whole motor giant system clicks out, and normal swimming under the finer control of the individual motor axons takes over."

In the next weeks, she notes her efforts and difficulties, particularly with the stimulating electrodes. Her analytical question remained unanswered, but it was a good technical apprenticeship, crowned with the occasional success, as on March 12: "Today I have on film clear evidence for facilitation of the motor giant axon at frequencies of 1/sec."

In spring 1971, Marder decided to try the stomatogastric ganglion. By that time, she and Brian Mulloney were firm friends, and he taught her the dissection and how to work on the finished preparation. Soon Mulloney had worked out a way to do it in three hours and then couldn't think why Maynard had made it so complicated. Marder now has scientists in her lab who can whip out the ganglion in forty-five minutes, but she says they all start, as Maynard and Selverston and others did, by not knowing what shortcuts can safely be taken without "losing" the preparation. It's only after many attempts that the fear of cutting precious nervous tissue away with the unwanted fatty tissue subsides. "So yes," she added, "I would imagine that after a year of doing them, Brian sped up by just *deciding* to go twice as fast."

On page 33 of her lab book, dated April 26, the top line reads, "First look at Panulirus Stomatogastric Ganglion. Today opened up a lobster and took out his stomach, laid it open and am using suction electrodes to explore the ganglion. Have one stimulating electrode, and one recording electrode." She went on to note, "Successfully recorded bursting ..." This means that she was seeing groups of several "spikes," or action potentials, in rapid succession on an oscilloscope (figure 2.2).

Many of the pages in this first notebook log the "poke it and see" trials that are invaluable for young scientists. Left to herself, Marder could try rough and ready means to explore.

After that one first look, Marder spent the next few months studying the cardiac ganglion and continued to try out various biochemical and electrophysiological techniques. In addition to this experimental work, she was reading and thinking furiously. In June 1971, she speculated on what might make a neuronal network more or less responsive, that is, set a system's "tone." Instead of changes in the responsiveness of an individual neuron, such as habituation and facilitation, that are due to patterns of stimulation, she was considering a more pervasive cause of change that might act on all the neurons in a network. Thinking about the large molecule assemblies that allow tiny sodium ions to move across a cell's membrane to maintain its electrical charge equilibrium, she wrote, "Could this be the place at which hormones interact, very subtly, to turn up or down the overall performance of the system without interfering at all with the actual phase relations and interactions of the system?" Astonishingly, she was already fumbling toward the idea of neuromodulation more than ten years before publishing her influential work on the phenomenon. Here, the question was posed rhetorically; at the time, she had no means of investigating it, nor did anyone else.

By September 1971, the first notebook was filled up. Marder had worked on nineteen preparations, a total that represents a great deal of painstaking work for a "hippie chick" (her description). Now it was time to commit to a thesis project. I suspect Marder had known for months that she would choose to work on the stomatogastric ganglion. Although the cardiac ganglion looks easier to study than the stomatogastric because it has nine neurons, not thirty, it only produces a single cycle, oscillating on and off, on and off. For research into neuronal networks, that's rather too simple. The stomatogastric ganglion, on the other hand, produces two distinct

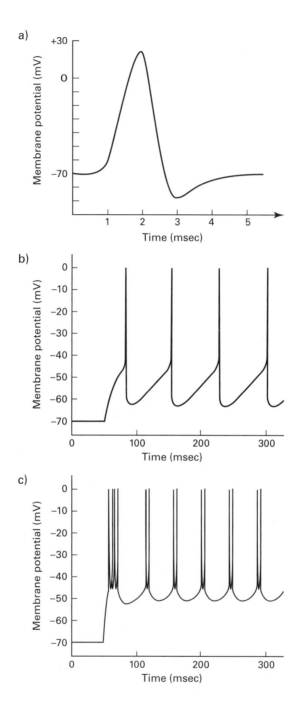

Figure 2.2

Representation of the waveforms associated with typical neuronal activity: (a) canonical action potential, (b) tonic firing, and (c) burst firing.

rhythms, the gastric and the pyloric, controlled by two overlapping sets of neurons. Don Maynard had already worked out the connections between the neurons running the pyloric rhythm and had identified the muscles they controlled. Working beside Marder in the lab, Allen Selverston and Brian Mulloney were using electrophysiology to map out the "wiring diagram" of the gastric circuit.

Marder was interested from the start in what substances would be found in the ganglion's neurons, what they used as neurotransmitters. She didn't want to study connections between neurons without knowing what neurotransmitter each one used; she was convinced the message was as important as the delivery route.

In a heroically overreaching moment, young Marder decided that, for her doctoral research, she would find out what *all* the neurotransmitters used in the stomatogastric ganglion were. She had realized that knowing them all would be fundamental to understanding the mechanisms of the ganglion and how it produced its rhythmic activity. The wiring diagram on its own would not be enough.

In her second lab notebook, in late October 1971, she set out a "Plan for Stomatogastric Ganglion stuff," which included grinding up nerve cells that had been identified according to the particular muscle they controlled and then testing the mash to find out what neurotransmitter they contained. Selverston called it weird pharmacology, but he let her get on with it. In fact, she had very little idea how to do any of it.

* * *

In the 1970s, identifying and describing the chemical substances that are used by neurons was difficult and time-consuming, but all the same it seems astonishing that only forty years ago so few neurotransmitters had been described. Part of the explanation may be that, through accidents of history, neuropharmacology got a late start; it was long thought that the only means of communication in the nervous system was electrical. That belief endured from 1791, when Luigi Galvani published the results of experiments using frogs' legs. After observing that an electric spark at the nerve leading to a muscle made that muscle twitch, he elaborated a theory of "animal electricity."

Generations of scientists followed this lead, which is unsurprising given that electricity was a nineteenth century passion. Of course, even at the beginning of the twentieth century, techniques for purifying, analyzing, and identifying chemical substances were few and laborious. So although it was obvious that cells both contained and were bathed in fluids, the composition and actions of the substances in those fluids were not known. Cells and tissue samples dissected from animals were kept alive in simple saline solutions. As for nerve cells, only in the early years of the twentieth century was it universally accepted that they were indeed distinct cells rather than a web-like network with animal electricity coursing through it. So it was perhaps natural that physiologists assumed that the electricity jumped the gap between each neuron and the next. What was known of chemicals in animal bodies seemed to rule out chemical messaging across the gap as much too slow for use in activating muscle, let alone the transmission of thought in the brain.

Many distinguished scientists were working on nerve cells and the nervous system in the first half of the twentieth century. In 1914, an Englishman, Henry Dale, made the important breakthrough. He isolated a substance called acetylcholine from a fungus and found that it slowed heart rate in frogs, implying that it or a similar substance might be used by the neurons controlling the heart. But acetylcholine is so unstable, so short-lived, that he could find no way of showing it was naturally present in animal tissue. His finding was much debated, and a long battle began between adherents to the electrical communication school and advocates of chemical communication at the synapse. The dust has long settled on an argument that had truth on both sides: some communication between neurons is in fact electrical (and an important example is found in the stomatogastric ganglion).

In 1926, Otto Loewi demonstrated "neurohumoral transmission" of acetylcholine at the synapse between neuron and muscle using frogs' hearts, but there was still no proof that the acetylcholine came from the nerve rather than from the heart muscle. However, this finding prompted Dale to take chemical communication seriously again, especially after 1929, when he found acetylcholine in animal tissue, proving at last that it was not specific to fungi. He picked up the subject energetically, and by the early 1930s, he had gone far enough to state that acetylcholine was indeed secreted by motor neurons onto skeletal muscle. Dale shared a Nobel Prize with Otto Loewi in 1936 for their work on what was then called "chemical neurotransmission."

In 1944, the eminent Harvard biologist George Howard Parker listed the terms in current use for substances that "serve as activating agents for other parts of the nervous system." They included "neurohumors," "transmitters," "neurohormones," and "neural chemical mediators." Dale preferred "transmitters," while Parker opted for "neurohumors." In addition to acetylcholine and adrenaline, they would have agreed on sympathine (renamed first noradrenaline, then norepinephrine) and the more speculative "intermedine," now lost to science except as a constituent of borage leaves.

Henry Dale's influence extended to the 1970s, when Marder was starting her research. Although the list of known neurotransmitters had by then grown to about ten, it was still generally repeated as a sort of doctrine, known as Dale's Principle or even Dale's Law, that one neuron could release only one neurotransmitter. By the late 1970s, that theory was disproved, and it is now known that a neuron can secrete more than one substance. In any case, the doctrine was a misunderstanding of Dale's position. What Dale actually said, which is subtly different, was that a neuron released the same transmitter from all of its terminals, even when it connected with many different target tissues.

Marder's project was hugely ambitious for a doctoral student. With the techniques of the early 1970s, there was no way she could have completed it in a reasonable amount of time for a doctorate. There are fourteen different types of neurons in the stomatogastric ganglion; she would have had to find each one's neurotransmitter. No one imagined then that dozens of neurotransmitters would eventually be found in the ganglion. I asked her if, when she set herself that goal and listed all the experiments required, she had imagined she could do it. With a rueful groan she answered, "I thought it would all be possible, yeah, totally."

Her motivation was not an obsessive fixation on completeness for its own sake. The stomatogastric ganglion was interesting because of its rhythmic output, and the overarching research question was which neurons or what neuronal property generated the rhythms. Like her supervisor, Marder wanted to find out how the whole ganglion operated, but she thought that neurotransmission was as big a part of the answer as the connectivity diagram. That diagram would tell you which neurons communicated with which others but not much more. Different neurons used different neurotransmitters, but why? Marder, knowing that nature is usually economical and doesn't waste much, was intrigued; if all these different substances existed, then there should be some rules for how they worked, and these rules would only be revealed by studying a whole system.

It was an entirely original goal. As she says, scientists are good at compartmentalizing; at that time, research on the known neurotransmitters was curiously fragmented, almost by geography. Sweden led the field in research on glutamate, England was known for acetylcholine studies, while much of the work on gamma-aminobutyric acid (known as GABA) was taking place in Boston and Los Angeles. No one seemed to be thinking about what would happen at the level of a single neuron receiving inputs of the multiple neurotransmitters found in its network. But Marder's insight, a rich one, was that because each neuron in a network had contacts with many others, it would probably be influenced by more than one neurotransmitter, and so it would be more meaningful to identify all the neurotransmitters in a system than to concentrate on just one substance.

Going further, she speculated that if each individual neuron could respond to multiple messages from many different transmitters, that would mean it must have receptors somewhere on its surface, at synapses or elsewhere, for each of those different substances. The proportion of different kinds of receptors would dictate a neuron's responses. If that proportion could change, and this was a wild conjecture, that would change the balance of the neuron's receptivity to different transmitters, and that might change the neuron's "tone" or might allow a neuron to "learn" to react in different ways. This formidable line of reasoning, untestable in 1971, became the basis for a decade of Marder's work. Nowadays, the role of receptors is studied in many labs, and elegant techniques have been developed. Marder's lab is currently using one called "laser uncaging" to map receptor distributions.

Sitting in her lab at Brandeis with the shelf of lab books to hand, I read her notes to herself, which are interspersed with the record of her experiments, and try to sort out the sequence of her thoughts. It's not easy because her thinking often ranged far from the project she had set herself; that project had to be tackled neuron by neuron and was mostly recorded with more mundane notes. In November 1971, Marder noted a list of the questions she was mulling over. It's a fascinating list, showing the clarity of thought and the attention to the natural behavior of an animal that are characteristic of her.

After a first straight question, her chosen research question, "What are the transmitters in the stomatogastric ganglion?", she asked, "Are both excitatory and inhibitory processes mediated by the same transmitter? Or, alternately, does a cell with dual functions have in fact two transmitters?" That subversive thought led on to, "How many transmitters is one cell responsive to?"

Then she wondered whether the different neurons in the ganglion differ in their responsiveness to various pharmacological agents, and went on to a fifth question: "From all of the above, is it possible to make any guesses about whether it is a useful model to think of a small nervous system as a circuit of equivalent units, or in fact do the units have to be individually characterized?" Wiring diagrams in effect treated nervous systems as circuits of equivalent units, the neurons. The prevailing theoretical framework was that neurons had all-or-nothing responses to excitation, on or off, so the important thing was the connection diagram. Marder already doubted that this was sufficient to explain the network's mechanisms without describing how each of the neurons worked, what neurotransmitter it used to signal, and what receptors governed its responses.

She followed this with more general musings: "A long shot—do animals that have been starved in any way modify their chemical responsiveness to pharmacological agents? Or is this perhaps not the quite right question? How to tell whether these cells actively change their surface properties in response to varying environmental conditions?" The neurotransmitter secreted from a neuron binds to molecular structures called receptors that are concentrated at the synapse of the cell it is signaling to; the synapse is the site for what one might call "direct messages." But there are also receptors and ion channels all over the cell membrane, and these interact with the many substances in the cell's environment. Changes in the number of these receptors, or in the relative numbers of different receptors, would change the cell's disposition to react. Here Marder is suggesting a complex feed-forward process.

The next question might as well have been science fiction at the time. In the 1970s, gamma-aminobutyric acid (GABA) was one of the most studied neurotransmitters, but little was known about receptors, let alone the answer to this: "Are chemically sensitive receptors inducible in any way? Can you get a cell with no responsiveness to, for example, GABA to become sensitive to GABA by putting it in culture in the presence of GABA in the bathing solution?" When she wrote that, the experimental methods available allowed only indirect clues about receptors. For example, in work just recently published from Paris, the American scientist JacSue Kehoe had revealed that there were three distinct receptors for acetylcholine by showing that there were three distinct changes in membrane potential with different timescales for its effects on the electrical properties of a neuron's membrane. Today, the structures of receptor molecular assemblies can be analyzed and visualized, but that relies on advances in genetics and crystallography that Marder could not have dreamed of in San Diego. She says she was always interested in receptors, "but I couldn't do anything about it."

In January 1972, she wrote, "I want eventually to be able to design experiments to test whether or not the surface properties of a neuron are constantly changing. Change during development or can be induced to change. Find out what kinds of receptors are on each cell. Do nerve cells have receptors for transmitters they don't see?" Neither she nor anyone else could answer those questions; most neuroscientists weren't even asking them. In this lab book, she could only ponder ways of visualizing the distribution of the receptors on a cell's surface before returning to the lobster's environment and the control of its gastric mill: "Possibility of modifying the patterned output by using different feeding schedules (hard food, soft food, frequent feeding, no feeding)."

* * *

The academic year September 1971 to August 1972, her third at San Diego, went by, and there is just one notebook for the whole period. At the bench, she was struggling to demonstrate unequivocally that *any* particular neurotransmitter was used by *any* one of the ganglion's neurons. It was generally accepted at the time that several criteria had to be met before a substance was proved to be the neurotransmitter of a particular neuron. Marder made her own list that she thought included the essentials of other people's lists and tried to conform to "my little bible" of proofs. One had to show that the neuron both contained and made the substance and secreted it in sufficient quantity to have an effect on the receiving cell, which could be a neuron or an "effector" such as a muscle. The receiving cell had to be able to respond to the substance, thus confirming that it had the appropriate receptors in its cell membrane. There had to be a mechanism for terminating the neurotransmitter's action, such as an enzyme in the gap between the neurons to break it down, or reuptake by the signaling neuron. Trying to prove all this for the neurotransmitters in each of the ganglion's fourteen types of neurons was, as I've said, a brave undertaking.

The initial step was to show that a particular identified neuron was able to synthesize a particular neurotransmitter. Her immediate problem was to distinguish and identify the cells so that she could find and recognize them in all of her preparations of the ganglion, by both visual inspection and their electrophysiological properties. This is not a trifling matter; crustacean neurons aren't color coded in the animal (unlike those of molluscs, which are helpfully pigmented in different shades of orange). I'm going to take the time here to outline the procedures Marder had to follow just to pick out her cells of interest.

First, catch your lobster. They were kept in the Scripps Institution's tanks in a locked shed almost on the shore. Seawater was pumped continuously from the ocean into the big concrete tanks, making an ideal healthy environment for the lobsters because the water wasn't recirculated, and its temperature was that of their natural habitat. These were perfect conditions, it seemed—until one summer, the pumps failed and everything in the tanks died. There were several inches of repulsive muck to be cleaned out. Marder and Mulloney had to put on rubber boots and wade into the stench.

Usually the lobsters thrived and were feisty and hard to catch. The grab has to be accurate, at the thorax, which is rigid. Spiny lobsters don't have claws, but a grab on the abdomen ends in cut hands as the rapid tail curl catches skin between the shell plates. They aren't called "spiny" for nothing: the sharp rasping spines on their antennae seem to be purpose-made to scratch unwitting graduate students. Mulloney and Marder used an old dish rack to reach down and try to scoop a lobster within reach for the grab. They got pretty good at it. But it was never easy for Marder, who is not tall; the rim of the tank was about five feet off the ground, and that meant she had to hoist herself up and balance on her stomach to reach in. Of course, once they had missed one lobster, none of them would come near the side of the tank, and, by the way, the tanks were enormous; each could hold two to three hundred animals. One impatient scientist would have to chase the lobsters with a long stick while the other tried to net them with the dish rack as though they were butterflies.

Their catch, usually a lobster each, had to be carried a mile up the steep hill to the campus. Mulloney remembers taking the bus with a lobster in a bucket, its antennae waving out of the top, but apparently unnoticed by the other passengers. Marder says, no, they went by car, putting each lobster in a brown paper bag on the back seat or on the floor, where the bags would tumble around as if possessed. Then she remembers that her car died in the last six months of her graduate work, and grins, yes, she probably made the bus trips too.

What follows here is a short and not too gruesome description of dissections I've watched in Marder's lab. The lobster is submerged in a bin of crushed ice to anesthetize it. There are few rules to be followed, nothing more exacting than in a restaurant kitchen, because the law does not count crustaceans as animals. An important next step is to put the tail in the lab freezer. Stomatogastric ganglion labs are popular in their departments; they have a constant supply of lobster tails (and claws if they're using Maine lobsters or crabs), and these can be used to foster good relations with colleagues, secretaries, and friends. Researchers go off them pretty quickly, as Brian Mulloney says, "After the first year you saturate and stop eating them." Marder doesn't eat them anymore either. "It would be like eating a colleague."

In dissection, the mandibles are removed and the carapace cut open. The esophagus and a fairly clearly delineated digestive system are cut out of the sloppy mass surrounding it. Under a microscope, the fatty and connective tissues are carefully picked away. Usually the preparation includes the esophageal and commissural ganglia and the stomatogastric ganglion itself, with the motor nerves running from it. These last contain the axons of the motor neurons. The somata (the cell bodies of the neurons) are distributed on the outer surface of the ganglion; inside is a dense tangle of axons and dendrites called the neuropil. The preparation is put in a glass dish, and fine pins are used to hold it in place on the base coat of clear elastomer. The ganglion looks like a blob of half-cooked egg white with the ghostly white bubbles of the cell bodies scattered over it.

The difficult trick is to keep a good length of the motor nerves that run from the ganglion to the pyloric and gastric muscles intact because they provide the means of identification of the neurons: matching the action potentials of a neuron in the ganglion with activity in the motor nerves. The ganglion and nerve fibers are contained in a sheath that protects them from the rubbing and mechanical wear of the muscular digestive action, which is robust: the stomach rolls up and over. Most of the sheath has to be removed from the ganglion so it can be accessed with electrodes. The bits of sheath still left attached are pinned away from the ganglion, tightly stretched out like the ropes of a tent to hold the ganglion flat and prevent it from flopping around. To isolate small stretches of nerve where an extracellular electrode may be used, the researcher makes rings of petroleum jelly, or "wells," in the elastomer base around them. Then the preparation is covered in saline solution, and the dish is moved onto the rig.

Ah, the rig, the focus and bane of an electrophysiologist's life. The rig has three sides and a ceiling to protect the setup from electrical interference, drafts, and jolts. It is just big enough to house a microscope, the micromanipulators, and various stands and supports for fluid leads, temperature probes, and any other pieces of kit required for the day's work. (See plate 1.)

Stomatogastric ganglion work often involves simultaneous recordings from many electrodes at the same time. These electrodes come in two flavors: intracellular ones, which record millivolt signals from the interior of a single neuron, and extracellular ones, which record microvolt signals from the motor nerves, each containing the axons of several motor neurons. On the rack next to the rig, a bank of amplifiers boosts the signals from the electrodes and transmits them to an oscilloscope or directly to a digitizer board and thence to the computer. Often an amplifier is connected to loudspeakers as well, and the blipping sound of action potentials or the buzz of their rhythmic bursting helps to identify different neurons by their firing patterns. In a stomatogastric ganglion lab, everyone recognizes those signature tunes.

The rig is festooned with wires and tubes, the rack beside it is packed full of electronic equipment, and sticky notes flutter in the air conditioning to identify bits of apparatus, list recommended settings and technicians' phone numbers, or warn against a variety of possible calamities. The researcher perches on a stool beside a computer, which is usually on some sort of trolley and looks as though it's on life support from the rack. The computer has, of course, replaced the cameras and trace printers of earlier years. (See plate 2.)

Measuring the changes in electrical charge across a neuron's membrane, which are minute, is the basic electrophysiological method for measuring the activity of that cell. It requires an electrode on each side of the membrane; the intracellular microelectrode penetrates into the cell while the reference electrode is in the fluid outside it. The microelectrode has to have a tip fine enough to just pierce a cell's membrane, which is about a millionth of a millimeter thick, without doing too much damage, but wide enough to allow contact between the fluid interior of the cell and the solution of charged ions contained in the electrode. Microelectrodes are made from fine tubes of glass, and the tip is made by heating the tube and pulling the two ends apart until it breaks. Microelectrodes are fragile, capricious, not shipped in bulk by laboratory suppliers, and not kept rolling around in drawers. You make your own when you need them.

The cell bodies lie round the dorsal side of the ganglion. Some of them can be recognized by size and position, but it's always vital to check; there's quite a lot of variation from preparation to preparation. Just look at the different arrangements of the blood vessels on the back of your left and right hands. Conveniently, stomatogastric ganglion motor neurons have relatively large somata, which are not spherical but more like fat frisbees (with diameters ranging from 25 to 120 µm), and can be successfully probed with a microelectrode almost every time, or even with several electrodes at once. For comparison, the tip of a microelectrode is about 0.5 µm in diameter.

The electrodes are fixed in a micromanipulator, and nowadays a microdrive is often used to position them, avoiding the inevitable wobbling of the human hand. But in the 1970s, the researcher, squinting anxiously down the microscope, would move the microelectrode to just touch the cell and then timidly give it a little tap, hoping to penetrate the membrane without damaging the cell. That was before the "buzz," as Marder remembers: "When did we all transition to buzzing? It would have been probably in the 1980s. Touching and then buzzing. I remember the first amplifiers had a capacitance button, but I don't know who figured out that you could use the capacitance compensation to actually do a buzz, and then somebody put a buzz button on an amplifier. But that was after people were already buzzing by just turning the cap button." That's electrophysiologist speak for you. The success of the final, infinitesimally small movement to just pierce the membrane still seems to require a dose of luck.

Once Marder had identified the cell body she wanted, the next step was to find out what neurotransmitter it synthesized, and this was a challenge because she wasn't working in a biochemistry lab and had to try out methods on her own. Electrophysiology is good for information about a cell's activity but not what substances are inside it. From January to June 1972, Marder struggled, most of the time with a technique called electrophoresis (no relation to electrophysiology), which I'm not going to describe because it isn't important in this story. There were a great many failed experiments, but her record of them is matter of fact, and I don't think it was a hard time for her. She had a boyfriend who was also a graduate student in Allen Selverstons's lab; he had arrived a year after her. With Mulloney, they rented a house that was one block away from the beach and biking distance from the lab if you could face the wicked hill. Marder and Mulloney played soccer in a mixed team, and they all did their share of partying and bodysurfing. She made friends easily, and her circle was certainly not limited to fellow biologists. She was also reading prodigiously, thinking, planning, talking, and especially listening to her colleagues. Selverston's lab was expanding, and new neuroscience labs were starting up on the campus.

Much of the recently published work she was reading used molluscs and crustaceans and built on the pioneering work of Edward Kravitz at Harvard in the 1960s. Marder was particularly impressed by his 1967 paper, coauthored with Otsuka and Potter, that began, "There is little doubt that gamma-aminobutyric acid (GABA) is the inhibitory transmitter compound at the lobster neuromuscular junction." So that was inhibition squared away then, but what about excitation? It was already known that, in mammals, motor neurons signaled with acetylcholine to activate skeletal muscles. But it was thought that glutamate was the excitatory neurotransmitter at crustacean neuromuscular junctions, at least the few that had been investigated, which were mostly in walking legs or claws. Of course, the motor neurons of the stomatogastric ganglion also work on muscles and make neuromuscular junctions with them, so it might be expected that they too would use glutamate. Marder made no prior assumptions based on these reports, however, and planned to test everything.

After she had tried to find GABA in single motor neurons, "Mystery—I haven't seen any GABA yet," she tested its action on all the neurons at once: "GABA in the bath turns the whole ganglion off." These were clues that GABA played no part in the neuromuscular communications of the ganglion but did affect its neurons. Marder turned her attention to acetylcholine.

In June 1972, she heard about Richard McCaman's lab, where researchers were successfully detecting the enzymes for synthesizing specific neurotransmitters in single cells. This was remarkable because the quantities of substances being analyzed were so small. McCaman was using a dependable assay for the enzyme that synthesizes acetylcholine, choline acetyltransferase. Its presence in a neuron would be good evidence that acetylcholine, that notoriously ephemeral molecule, was being made by the neuron. The assay was radiometric, using a radioactive form of one of the precursor molecules of the enzyme. But how was it actually done? McCaman's lab was just two hours up the road in Los Angeles. So she rang him up and asked to learn how to do his assays. On July 2, she wrote in her lab book, "Went to see Richard McCaman. Wow!"

McCaman taught her the assay and showed her his distinctive way of dissecting out a cell body, under a microscope of course, using ethylene glycol freeze substitution to make it firm enough to pick up with forceps. Then the cell body could be put on the tip of a fine glass rod, and by sliding a glass tube over the rod, the cell could be deposited at the bottom of the tube. This positioning was important because of the minute quantities of chemicals that would be used in the analysis. On July 3, she wrote, "Tonight, I used McCaman's method of using a glass needle pulled out under the microscope and transferred single cells that way. It's easy."

After a few days, McCaman suggested she should stop the long commute to Los Angeles and work with someone near her own campus. Dave Shubert, who had been trained in the assays by McCaman, was at the Salk Institute in San Diego. The lab at Salk included several young men destined to become distinguished scientists: Jim Patrick, Steve Heinemann, Yoshi Kidokoro, and Jon Lindstrom. She is still impressed by their generosity and describes them as model colleagues. "They said, 'Well, that's nuts to set up biochemistry in Selverston's lab, just come over here and work in our lab.' These were the good old days, right? They gave me carte blanche in their lab, and all the biochemistry I did from then on worked!" It is easy to imagine that, in return, her enthusiasm and insatiable appetite for reading research papers made her an interesting person to have around the lab, something of a walking science newscast.

So Marder carried her single cells in a freezer box a mile down the road to the Salk. By the end of July, her results showed that at least four of the eleven types of motor neurons in the stomatogastric ganglion contained the enzyme for synthesizing acetylcholine.

* * *

By September, Marder had started on the next stage of proof required by her little bible and was trying various methods to demonstrate the effect of acetylcholine and other transmitters on the whole ganglion and on single identified cells from it. She made many false starts, and what Marder now calls "crazy assays" that didn't seem to work. She began by "just dumping" each of the known neurotransmitters onto the ganglion to see what it reacted to. Her assumption was that if she saw a result, it meant that at least some of the neurons were responsive to that substance. But all of the neurotransmitters affected the ganglion. They altered its rhythmic motor pattern, each in a different way, and Marder realized that this was not going to tell her what she wanted to know. There was no way to analyze these puzzling results, and she had no conceptual framework to interpret them.

She took her problems and ideas to Mulloney, her stalwart friend and mentor, and talked them over with the Salk group and with Nicholas Spitzer, who had joined the UCSD faculty. Marder is rarely isolated; her easy sociability and intense interest in biology propel her quickly into discussion with every like-minded scientist she meets. She had long conversations with a young assistant professor from Oregon, David Barker, when he visited San Diego, and she was thrilled that he thought her results, although confusing, were nonetheless promising. He recognized them as clues that the neurotranmitters she was putting on the ganglion might not be acting as transmitters at synapses but more like hormones in a diffuse manner. In the living animal, these substances might be present in hemolymph or secreted into the neuropil by the input nerve to the ganglion. He too had predicted that neurons received multiple influences as well as direct neuron-to-neuron signals. Later she would go to his lab for her first postdoctoral experience.

Marder's scientific horizons were expanding, and she felt she would soon have had enough of San Diego and graduate school. In March 1973, a new note of determination is noticeable in her lab book: "Need to get 80 experiments to *work* within 1 year. That means, need to get 2 experiments to work/week!!!!" The pace quickened.

One of the four types of neurons that Marder had shown contained the enzyme for acetylcholine synthesis was the pyloric dilator (PD). Each stomatogastric ganglion contains two pyloric dilator motor neurons that contact the dorsal dilator muscle of the lobster's pylorus, where ground-up food passes from the stomach. As we have seen, glutamate had been found to be the excitatory neurotransmitter at some crustacean neuromuscular junctions, so this evidence of acetylcholine was an unexpected and intriguing result. By itself, however, it was far from enough.

The question that now had to be answered was whether acetylcholine activated the muscles these motor neurons controlled. If it did, then that would be another indication that it might be the neurotranmitter used by the motor neuron signaling to that particular muscle. Knowing that the pyloric dilator motor neurons commanded the dorsal dilator muscles, she realized she could use those muscles to find out.

She solved the three-dimensional puzzle of dissecting out a muscle with its PD nerve attachment and went to work. Two of the key measurements in electrophysiology are the voltage difference between the inside and the outside of a cell, called the membrane potential and measured as voltage, and the resistance of the cell, which is ascertained by injecting known amounts of current and measuring the resulting voltage deflection. From these, she would plot a graph of the values measured, current (for which "i" is the symbol) against voltage (v), giving what is known as an i.v. curve. The current depends on the number of ions flowing through channels in the membrane, which are expected to open in the presence of the right neurotransmitter.

Dripping carbachol (a readily available—and stable—surrogate for acetylcholine) onto the muscles and expecting to measure small changes in the muscle cells' electrical properties, Marder was frustrated by the muscle movements. However good her dissection, however good the initial baseline results after carefully putting the two electrodes in the muscle fiber, when she added carbachol, the muscle contracted and the electrodes flew out. On the fourth or fifth day in a row, exasperated, she went screaming down the hall:

"Every time I put the goddamn carbachol in, the goddamn muscle moves."

Halfway down the hall she stopped:

"Wait! Every time I put the transmitter on the muscle, the muscle moves! How can I be so stupid? I've already got the answer!"

So she walked into Selverston's office and said,

"Al, I've been trying to do i.v. curves so I could measure changes in conductance, but the muscle moves and that's because I'm using its excitatory transmitter."

"Well, you can't publish saying I looked in the microscope and the muscle moved. Hook the goddamn muscle up to a tension transducer and measure the movement."

That floored her because she had no idea what a tension transducer was. Selverston fished one out of a drawer, saying, "Here!"

In October 1973, her sixth lab book records a successful experiment activating the dorsal dilator muscle with carbachol and measuring the strength of the muscle's contractions. No one should underestimate the time-consuming laboriousness of these experiments. The tension was shown on an oscilloscope, and to have a record of it, a camera was set up in front of the screen. But tension recordings are slow, so Marder used to stop the sweep of the oscilloscope and just film the spot going up and down. Because the camera was set at a really slow speed, in effect slowly moving the film past the moving spot, it had to be done in an almost completely dark room. Her handwriting in the lab book is big and scrappy. There were plenty of steps to go wrong. The last straw would be when the muscle contracted too much and she lost the trace off the scale. But she managed to get all the data for her thesis that way, sometimes having to redo a successful experiment just to get a film of the trace good enough to publish. As Marder says, "The kids of today don't realize—they just push the button on a computer."

* * *

By the spring of 1974, the experiments she recorded were dotting the i's, getting better cleaner traces, and checking everything while she started to put her thesis together. Selverston said, "I thought you were going to do more." Marder said, "No, I've finished."

She was aware that her main result was significant and decided to write it up and submit it to the prestigious international journal, *Nature*.

After that, she turned her attention to a few final experiments. Brian Mulloney was now married and had moved to the Davis campus as an assistant professor. Marder wrote to him in July: "Tested *all* the muscles on the stomach (more or less) for sensitivity to acetylcholine ... and neurons, everything is acetylcholine sensitive except IC, LP, PY. ... I hate this terminology. ... Not much else happening except I figured out how to serve hard in tennis." The terminology used for the neurons, nerves, and muscles of the stomatogastric ganglion is truly byzantine. There is a list of the names of stomatogastric ganglion neurons in this book's Glossary. Fortunately, only a few of them have starring roles in the story.

Then her thesis had to be written in the customary, monumentally time-wasting form. She wrote again to Mulloney in August: "Things here are very quiet except I'm reaching a peak of irrationality with my only sane moments when I'm playing soccer or tennis. I've been playing Tues, Thurs soccer and have been tolerated, if not accepted. (We have the number of women up to five, so it's better, but there are still some hard-core mcps* who will not pass the ball to a woman even if they are bound to lose it if they don't.) ... I've started writing and am still finishing one last experiment. ... I'm just tired of being here. I want to be done more than anything else. I want to be done + go somewhere else and do different things. My rig is showing the strain—I've got a chart recorder, three stimulators hanging by wires, and have five manipulators poised around one little dish; would make your heart sink to see the chaos. Yesterday out of desperation I read all of Trollope's *The Eustace Diamonds*—all 700 pages. ... I'm determined to write my thesis in a week."

Nature published her report in October 1974, but any pleasure Marder might have had in the achievement was soured by the copy editors' horrible mistake after she had signed off on the proofs, for some reason substituting "acetylcholine transferase" for "choline acetyltransferase." After many aggrieved complaints, *Nature* finally published an erratum in January 1975, but without the apology that would have placated Marder.

In the paper, she showed that the pyloric dilator neurons synthesize choline acetyltransferase, that the dorsal dilator muscle they control responds to acetylcholine, and that its responses are blocked by antagonists of acetylcholine. She ended it with a clear, confident summary:

> It is now clear that the PD and LP (lateral pyloric) cells utilise different neurotransmitters. Furthermore, it is difficult to argue that L-glutamate is the neurotransmitter at the PD-dorsal dilator synapse, which clearly points to the dangers in making sweeping generalisations about 'the arthropod excitatory neuromuscular transmitter.' It is interesting that there is no known peripheral inhibition in the stomatogastric system, while many arthropod muscles receiving L-glutamate excitation also receive inhibitory innervation. It is also important that the PD and LP neurones make inhibitory synaptic connections within the stomatogastric ganglion.

This conclusion is already in classic Marder style. She was, of course, contradicting earlier hypotheses held by well-known older scientists and

*A younger generation may not recognize this acronym: mcp stands for male chauvinist pig.

didn't hesitate to emphasize that. I'm amused by the cheeky way she chose to express the put-down. But then, immediately after that, although not at all in a conciliatory way, she mentioned two special factors that might have some bearing on the case.

Her final statement is an exact summary of her findings, limited to what she could claim on her evidence:

> There appears to be heterogeneity in excitatory neuromuscular transmitters among the neurones within this one crustacean ganglion, and some of these neurones appear to make cholinergic neuromuscular junctions.

Her thesis was different from the *Nature* paper; not just a great deal longer, but, conforming to academic practice, covering a wider range of enquiry. It ended with seven questions about the neurotransmitters in the stomatogastric ganglion, questions she had been turning over in her mind throughout her time in Selverston's lab. I'm setting them out here because they show how her brave speculation of 1971 had crystallized by 1974 into a more systematic approach. She could now pick out the questions that linked the effect of neurotransmitters (and other substances from different sources) to the properties of the individual neurons and thence to the circuit as a whole. Within a decade, she would be able to answer the first five questions, but by then she had added many more.

"1 What is/are the neuromuscular transmitters released by the LP, PY and IC neurons?" These were the motor neurons for which she had only negative evidence: they didn't seem to synthesize acetylcholine.

"2 Is acetylcholine the neurotransmitter released by the PD neuron at its synapses within the ganglion?" One of the pyloric dilator neuron's two sets of synaptic connections is made outside the ganglion by the axons that leave it in the nerve fibers to contact muscles; the other is made inside the neuropil with other neurons of the ganglion. Marder had positive evidence that the PD neurons used, or at least synthesized, acetylcholine. Acetylcholine could now be expected to be an excitatory neurotransmitter at the neuromuscular junction, but the PD neuron's synapses within the neuropil were likely to be inhibitory. Marder knew of a very recent publication by Hanley et al. showing both choline acetyltransferase and a precursor of dopamine and serotonin in a single snail neuron. It wasn't proved that the snail neuron released either dopamine or serotonin at synapses, but it was a tantalizing clue. If the PD neurons released a second neurotransmitter as well as acetylcholine, it would be contrary to Dale's Law. She was very aware that the answer to this question might break new ground and she was keen to attempt it herself. "It is all the more important to do this work, since there are few if any cases in which a neuron with two such different sets of known synaptic connections has been studied in terms of the neurotransmitters involved."

"3 How many different classes of acetylcholine receptors are there?" This question was followed by a page discussing JacSue Kehoe's work showing three types of response to acetylcholine in *Aplysia*, the sea slug.

"4 What kinds of neurotransmitters are released by inputs to the stomatogastric ganglion?" This referred to the "descending" nerves from the esophageal and commissural ganglia that were being studied by David Barker's group at the University of Oregon. There were already hints that these neurotransmitters were different from the ones the neurons of the stomatogastric ganglion itself were thought to use.

"5 Do cell somata have receptors to transmitters other than those released onto them at presynaptic sites?" Marder had suggested that, as the connections among the neurons of the stomatogastric ganglion had been established in the wiring diagrams, a neuron might be found that received all its synaptic input from neurons with known transmitters. Such a cell could be tested for receptors on the cell body for other transmitters; if they were found, it would mean that the cell was able to respond to influences other than what one might call the direct messaging of the circuit.

"6 Is there a fixed distribution of receptor molecules on the surface of identified somata?" Marder was interested in the possibility that a cell's level of receptivity might be modifiable, but that would depend on a changeable distribution of different kinds of receptors.

The seventh and final question was impressively all encompassing, especially at the end of a thesis on detailed work in a small system. It's what I'd call "vintage Marder": "What possible generalizations can be made about the relationship between the neurotransmitters used in the neurons of a complete nervous system and the functional relationships of these neurons? This being the major issue about the way nervous systems utilize different neurotransmitter substances, and still unanswered."

At the end of the year, she wrote to Mulloney, "I'm going to send you a copy of my thesis as soon as I can get it made. I finished in San Diego really quickly, I think I blew Al's mind, he told someone he'd be surprised if I did it before February, and I did it on Dec 5. However, I did that by leaving things undone, like making extra copies. You would have been proud of me at my thesis talk. I was just a *tad* nervous, but I vowed that everyone would understand it all, and they all did. … Of course, it's such a nice little story, that it's easy. Al had a party for me that night, and I was really happy and proud, and McCaman got soused. … Then I split."

3 Lobster Lore

At this point in the book, it's time to describe Marder's model preparation in more detail. Lobsters and crabs, crayfish and shrimps, are crustaceans. They all have stomatogastric ganglia, although obviously of different sizes and shapes. They have shells, but although the shell is tough, it is not as rigid as those of other crustaceans such as mussels and oysters. The lobster's body is segmented, but all the head and thorax segments are fused together and immobile. The six segments of the abdomen are flexible. They're popularly referred to as the "tail" and popularly eaten. The last segment is the real tail and bears the tail fan. Lobsters are called "decapods" because they have five pairs of walking legs on the last five segments of the thorax; in some species, the first pair has become claws. All the other segments carry pairs of appendages too, although smaller, the swimmerets on the abdomen, for example. Spiny lobsters, such as *Panulirus interruptus*, have impressive antennae but not the alarming claws that cartoonists love—*Homarus* has those. (See plate 3.)

Lobsters don't do lunch; they are mostly nocturnal, spending daytime in crevices and burrows or under wrecked ships. When a lobster catches something edible, a choice worm or clam, it stuffs the prey into its mandibles for preliminary crushing and swallowing. The mandibles are on the fourth segment in the head region, and the following five segments carry "maxillae" and "maxillipeds" that are used to scrape food into the mandible. The chunks of food pass along the esophagus to be collected in the cardiac sac of the stomach. Bit by bit, they are moved on to the gastric mill, where cartilaginous ossicles, the "teeth," slowly chew and shred the lumps. (See plate 4.) Then the cardiopyloric valve opens so that the particles of food can pass into the pylorus, where they are filtered for size and further digested before being passed to the intestine (see figure 3.1).

The motions necessary for moving and grinding food particles are provided by muscles attached to the shell of the thorax and the wall of the

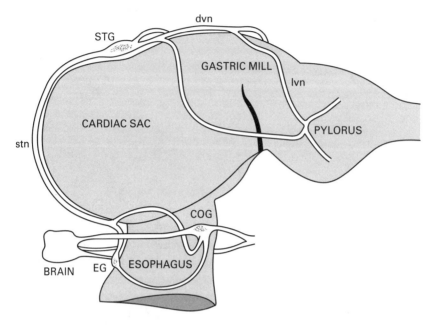

Figure 3.1
The three regions of the lobster's stomach and the esophagus showing the four ganglia of the stomatogastric nervous system. This diagram corresponds to the spiny lobster, *Panulirus interruptus*. EG, esophageal ganglion; COG, commissural ganglion (one each side); STG, stomatogastric ganglion; dvn, dorsal ventricular nerve; lvn, lateral ventricular nerve; stn, stomatogastric nerve.

stomach. When these muscles contract, they pull the walls of the stomach toward the shell, and this dilates the stomach or the cardiopyloric valve. They are termed "extrinsic" muscles because they attach to a part of the anatomy other than the stomach. The "intrinsic" muscles are attached at both ends to the same organ, either stomach or pylorus, so when these muscles contract, they squeeze the ossicles together or close the cardiopyloric valve. All the muscles are activated by the excitatory output of motor neurons that contact them at neuromuscular junctions. There are more muscles (about forty) than there are motor neurons (about twenty-three); each motor neuron contacts more than one muscle. The cell bodies of the motor neurons are part of the stomatogastric ganglion, itself part of the stomatogastric nervous system. It is housed inside an artery that runs over the top of the stomach (see figure 2.1 in chapter 2).

A group of neurons with some common function is called a ganglion. In invertebrates, the cell bodies are situated around and sometimes inside the dense tangle of their axons and dendrites called the neuropil. This arrangement allows neurons to make numerous and diverse contacts with each other and to do so in a shared environment—two important points in the organization of neuronal networks. The multiple contacts favor the establishment of complex circuitry, and in the shared environment all the ganglion's neurons receive similar information about the rest of the animal and its physiological state. The emergence of ganglia was an important step in the evolution of nervous systems, providing conditions conducive to the rise of complex behaviors and leading eventually to brains. The lobster, however, exhibits intellectual pretension and achievement only in Wonderland. Naturally, Marder has never bothered with its brain. In any case, that word would have to be in inverted commas, "brain," because it is just a large ganglion forward of the stomatogastric ganglion. It receives sensory information from the eyes and from mechanical and olfactory receptors on the head and the antennae. It controls several other ganglia posterior to it.

The stomatogastric nervous system includes four ganglia: a single esophageal ganglion of about eighteen neurons, two symmetrical commissural ganglia with about four hundred neurons in each, and the stomatogastric ganglion with about thirty. Together they control the rhythmic actions necessary for the swallowing, chewing, filtering, and transport of food, from the esophagus to the cardiac sac, gastric mill, and pylorus. The movement of food through the esophagus and the cardiac sac is predominantly controlled by neurons in the esophageal and commissural ganglia, which are only occasionally mentioned in this book. The neurons of the stomatogastric ganglion control the two sets of muscles, intrinsic and extrinsic, described above.

The principal input to the stomatogastric ganglion is the stomatogastric nerve from the esophageal ganglion, which is joined by branches from the commissural ganglia; it runs into the neuropil of the stomatogastric ganglion. In addition to axons from the neurons of the esophageal and commissural ganglia, it probably contains some axons from neurons outside the stomatogastric nervous system. It is the source of many of the neuromodulatory substances brought into that shared environment, and there is, to this day, no final count of those substances. Marder has often declared that no one can completely understand the workings of the stomatogastric ganglion until all of them are known and their effects on the ganglion described.

The neural circuits in the stomatogastric nervous system generate four rhythms: the esophageal, the cardiac sac, the gastric mill, and the pyloric.

They can run simultaneously, but each rhythm has a characteristic cycle period. The cardiac sac cycle is the slowest, ranging from about half a minute to two minutes. The pyloric rhythm is the fastest, cycling about once per second. The gastric mill rhythm cycles more slowly than the pyloric, with a frequency in the five- to ten-second range, but they have some influence on each other. These rhythms are shaped by bursts of action potentials in different types of neurons that alternate with each other. For example, the pyloric rhythm is triphasic: each burst by the two pyloric dilator neurons is followed by the lateral pyloric's burst and then the pyloric neurons. It is perhaps counterintuitive, but these bursts inhibit the neurons they affect, so that they, in turn, burst on release from this inhibition. All the chemical synapses between neurons of the pyloric circuit are inhibitory, although most of these neurons are also motor neurons and form excitatory connections with muscles (see figure 3.2).

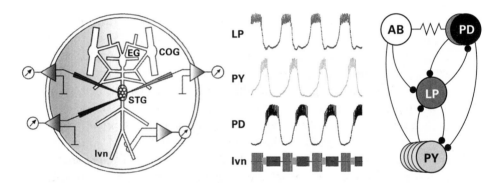

Figure 3.2
Electrophysiological work on the stomatogastric ganglion. The left-hand panel indicates the recording method using intracellular microelectrodes inserted in three different types of neuron that contribute to the pyloric rhythm and extracellular recording from an axon of the lateral ventricular nerve (lvn). COG, commissural ganglion; EG, esophageal ganglion; STG, stomatogastric ganglion. The center panel shows the three phases of the pyloric rhythm: LP-PY-PD. Note that no trace is shown for the AB, which bursts in synchrony with the two PDs. The bottom trace is recorded extracellularly from the lateral ventricular nerve, which contains axons from the LP, PY, and PD neurons and contacts muscles. The right-hand panel shows the connections between neurons of the pyloric circuit (black circles indicate inhibitory synaptic connections, the resistor symbol indicates electrical coupling). AB, anterior burster; PD, pyloric dilator; LP, lateral pyloric; PY, pyloric; lvn, lateral ventricular nerve. (Credit: the Marder lab.)

The cell bodies are on the periphery of the ganglion, and each sends a fine fiber, a neurite, into the ganglion, where it branches profusely. Unlike plant roots, to which neurites are often compared, they can branch to form thicker fibers rather than thinner as they go further from the somata. The motor neurons of the stomatogastric ganglion are unusual in that they both activate muscles and also participate, through their many contacts in the neuropil, in generating and maintaining the rhythms. The astonishing extent of any one neuron's communications can be seen in the messy knot of its many processes, interwoven with those of other neurons in the neuropil. The neuron's single long axon emerges from the neuropil to join the nerve toward the muscles. Dye-fills of single stomatogastric ganglion neurons show the complicated branching of the neurites. Apart from a vague similarity on visual inspection, no defining patterns for different cell types can be described (see figure 3.3).

So far it has been impossible to associate any particular zones in the neuropil with a functional specialization, although the finer branches of all cell types seem to be the site of synaptic connectivity and are more peripheral, while the thicker branches and axons are found in the core. The ganglion does not seem to be organized into different areas for different types of cell, whereas in many invertebrate ganglia each neuron projects only to certain parts of the neuropil.

The axons from each of the motor neurons in the ganglion branch again as they travel in the nerve that carries them to the muscles, and different branches from the same motor neuron can contact different muscles. Conversely, a single muscle can receive input from more than one motor neuron. This complicated innervation allows a subtle control of movement.

As a rule of thumb, it is usually accepted that there are about thirty neurons in the stomatogastric ganglion, of which twenty-three are motor neurons and the other seven are "interneurons" that do not contact muscle. However, in experimental work, slightly different numbers of neurons have been found in different species and, more confusingly, in different preparations from the same species. In Marder's lab, Dirk Bucher and Christian Johnson set out to fully describe the stomatogastric ganglion of the American lobster, *Homarus americanus*, and found a few surprises. Until their excellent and revelatory paper was published in 2007, and despite the extensive body of work on these crustaceans, the various anatomical aspects of the stomatogastric ganglion had only been studied separately; no one had put the whole picture together for a single species. They found that American lobsters can have variable numbers of stomatogastric

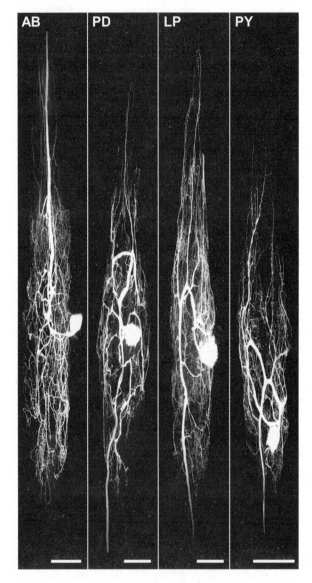

Figure 3.3
Dye-filled neurons of the pyloric circuit in a maximum intensity projection from confocal microscope scanning. The fine neurites extend throughout the neuropil and thus indicate the contour of the stomatogastric ganglion. The anterior burster (AB) is not a motor neuron and has no axon leaving the ganglion. PD, pyloric dilator; LP, lateral pyloric; PY, pyloric. Scale bars: 100 μm. (From Bucher et al., 2007.)

ganglion neurons: the total can be between twenty nine and thirty two, although most have thirty one. Among the thirty-one neurons, six remain mysterious because they don't seem to do anything, just receiving notice of activity in the esophageal ganglion. The numbers of the other types of neurons don't vary except for the pyloric (PY) and gastric mill (GM) neurons. These two are a real puzzle. In each lobster, Bucher and Johnson found from three to seven PYs and from six to nine GMs, with the total of the two always between ten and thirteen. Thus, the more PYs the lobster has, the fewer GMs and, of course, vice versa. This is curious because PYs and GMs do not do the same thing, are not even in the same circuit, and so can't be compensating for each other. What can the relationship be? Their cell bodies are positioned close to each other and always toward the rear of the ganglion, but they reach different muscles in different parts of the stomach using different neurotransmitters. The best guess at the moment is that they share a developmental path in the earliest stages of the animal's life, after which their differentiation is not tightly controlled (see figure 3.4).

Researchers in the United States habitually work on stomatogastric ganglia from the Pacific spiny lobster *Panulirus interruptus*, the Maine lobster *Homarus americanus*, and two crab species, the Jonah crab, *Cancer borealis*, and the rock crab, *C. irroratus*. Elsewhere, comparable work is done on related species found locally. When Marder opened her own lab, she initially preferred to work with the Pacific spiny lobster she had started with in San Diego. Soon, she began to use crabs to train students because they were cheaper and because "they have nice muscles." As the lab gained experience with them, crabs became increasingly a full alternative to the lobster. As her lab stacked up interesting results with crabs, other researchers started to adopt them too. Nowadays her lab uses mainly the Jonah crab; the rock crabs are no longer seen in Boston fish markets, possibly because the warming waters have driven them north. But in the 1980s, the lab was routinely carrying both of those crabs as well as the spiny lobster.

Using different species seemed the practical thing to do at the time. Although they all have a stomatogastric ganglion with the same or similar cell types and connections, and all generate the same type of motor patterns, the structure of the ganglion differs markedly—as you would expect when you consider the different body shapes of the animals. Even at a casual glance in a microscope, the *Homarus* stomatogastric ganglion is elongated (often called "spindle-shaped") compared with that of the crab (see figure 3.5).

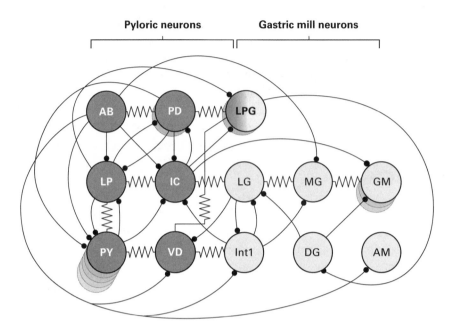

Figure 3.4
Multiple pyloric (PY) and gastric mill (GM) neurons shown in a diagram of the connections between neurons of the stomatogastric ganglion in *Cancer borealis*. Inhibitory synaptic connections end in black circles, and electrical coupling is indicated by
resistor symbols. Pyloric neurons: AB, anterior burster; IC, inferior cardiac; LP, lateral
pyloric; PD, pyloric dilator; PY, pyloric; VD, ventricular dilator. Gastric mill neurons:
AM, anterior median; DG, dorsal gastric; GM, gastric mill; LG, lateral gastric; LPG,
lateral posterior gastric; MG, median gastric. (Credit: the Marder lab.)

So much was always obvious, but gradually it became clear that there
were traps hidden in the assumption that the neurophysiology of the ganglion would be the same in all species. It does the same job for the animal,
which is to activate the muscles of the stomach and pylorus, but the devil
is in the details, and unwary researchers can be caught out by, say, a well-
identified incoming sensory or modulatory neuron that uses a different
neurotransmitter in different species, although carrying out the same function. Moreover, the different species can even have stomatogastric ganglion
neurons of the same cell type that respond to different descending modulatory inputs. This can make it hard to compare research results or, worse,
can lead a researcher to make assumptions that may hold for some species
but not the one under the microscope. Marder told me that the major connections seem to be identical, completely conserved throughout evolution

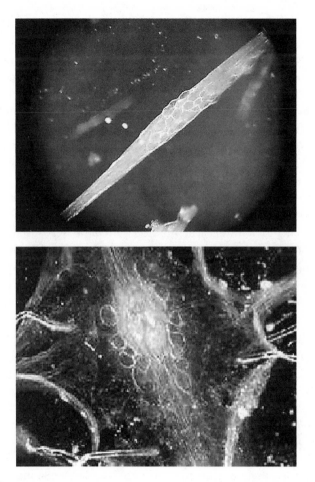

Figure 3.5
Stomatogastric ganglia under the microscope. Upper: *Homarus americanus*. Lower: *Cancer borealis*. (Credit: the Marder lab.)

in the different species. The least conserved features are in the descending pathways, the neuromodulatory complement, and cotransmission systems (as well as its principal neuromodulatory substance, a neuron may release other small molecules). "That's all over the place," she said. "And take the phase relationship in the pyloric rhythm: in *Homarus* the pyloric neurons fire into the beginning of the pyloric dilators' burst and they don't do that in *Panulirus* at all."

I asked her what these differences meant in the lab, in practice, working with these different species. "We've learned how to sort out what'll likely be a species problem and what isn't. The tricky stuff—and it's really annoying—is not going from *Panulirus* to *Homarus*—you know you're in another species, what's nasty is going from *Homarus gammarus* to *Homarus americanus* or from *Cancer pagurus* to *Cancer borealis* to *Cancer irroratus* because those are so closely related you're tempted to think they're going to be the same, but they still have differences. The structure of the crabs' stomachs is about the same, but the different species have different behaviors. They've evolved to inhabit different eco-niches so there has been some flipping of the modulatory inputs. *Cancer irroratus* are nasty aggressive animals; *Cancer borealis* goes passive on you."

The danger of extrapolating from one species to another is evident and certainly not exclusive to research on the crustacean stomatogastric ganglion. Similar problems exist in work on vertebrate nervous systems and are too often ignored. In fact, because so few of the myriad neurons in vertebrate circuits have been identified and characterized in any species, the danger is even greater. In species where gene expression can be examined, a further complication is sometimes found: similar anatomy, organs, or mechanisms in different species may turn out to have little relationship to each other. Among other factors, different evolutionary pathways or developmental progenitor cells may be behind these unexpected findings. More surprises may await the stomatogastric ganglion labs when gene expression can be fully analyzed for all the species they study.

In Marder's lab, researchers find their way to the species that seems appropriate for what they want to do. While she was a postdoc in the lab, Marie Goeritz made some instructive and fascinating digital "reconstructions" of dye-filled stomatogastric ganglion neurons. She found the crab stomatogastric ganglion easier to work with: "It's flat. It's so much easier to record from the cells without damaging them. The lobsters, both *Homarus* and *Panulirus*, are more three-dimensional; they flop around more so you have to pin them down very tightly and even then the cells are not lined up as nicely as they are in the crab. I was trying to look at the projections into the neuropil and that's one thing that's so beautiful with the crab: the neuropil is not covered with cell bodies. In lobsters, the neuropil has some cell bodies on top of it so to visualize the neuropil, you have to deal with those cells." (See plate 5.)

It took Goeritz months of work to get a single good photo. With special equipment to image deeply into tissue, she could produce tiled scans that then had to be stitched together. Ted Brookings, also a postdoc in the lab, wrote a program for stitching and optimization, and the lab invested in expensive image analysis software. Each preparation took at least an hour to dissect out the stomatogastric ganglion and get it ready in the dish. The cells could be identified and filled in about thirty minutes. The scanning of the dye-filled neurons took at least ten hours, sometimes twice that. Then the scans had to be reconstructed, and that, says Goeritz, is excruciatingly detailed work: "If you were to work full time on the computer it would be a week or two, but you go nuts if you do that." The lab formed a roster of undergraduates working a few hours at a time, and Goeritz estimates they ended up completing about a cell a month. The lab has now abandoned these full-volume reconstructions in favor of "skeleton" reconstructions that take much less time.

* * *

The stomatogastric ganglion does have some unbeatable advantages as a model preparation. It is especially convenient to have a model with its own intrinsic activity that does not depend on the researcher inventing a stimulus—itself a variable that has to be checked at every procedure. As you know by now, it has helpfully large neurons and so few of them that they can be named and consistently identified in preparations. There are few interneurons, which are smaller; most of the neurons are motor neurons producing muscle movement directly. Thus, activity in the nerves from the ganglion to the muscles can be used to monitor the output of the ganglion's neurons.

Unfortunately, the stomatogastric ganglion resides in crustaceans, none of which are ideal model animals for laboratory research. What's wrong with them? What is a good model? In the 1920s the Danish physiologist August Krogh gave the best-known answer to that question. He said that for any particular type of physiological problem, researchers should look for the animal in which it can be most conveniently studied. He went on to tell the story of his mentor, Christian Bohr, who "was interested in the respiratory mechanism of the lung and devised a method of studying the exchange through each lung separately; he found that a certain kind of tortoise possessed a trachea dividing into the main bronchi high up in the neck, and we used to say as a laboratory joke that this animal had been created expressly for the purposes of respiration physiology." Naturally, no one would choose crustaceans to study some simple attribute. However, the stomatogastric ganglion provides a complex but describable network that would be hard to match in other animals.

Anything with a carapace is manifestly inconvenient because you have to open up the shell to investigate; something soft and floppy like the sea slug has many advantages. But the dissection technique outlined in chapter

2 easily reveals the ganglion and its attachments, which are far more accessible to the researcher than ganglia in most vertebrate preparations. The main problems with the stomatogastric ganglion lie elsewhere.

Biology is no longer solely a descriptive science; the field expects research results to be analyzed statistically, and this means doing the same experimental procedure on many, many members of a particular species. Some of the more exotic and larger species that used to be studied have been replaced in the lab for this reason.

Biology is now wedded to genetics. Until the last few years, when new techniques of genetic manipulation on mature animals were introduced, several generations of an animal had to be studied to follow the effects of genetic modification. This consideration increased the importance in a model animal of a short lifespan, or at least a short time to reproductive maturity. So a convenient model species is one that can be bred quickly and housed relatively inexpensively; the stellar performers are the fruit fly with a generational turnaround of less than two weeks, massive hatchings, and a miniature appetite, and the tiny nematode worm, *C. elegans*, for similar reasons. Lobsters and crabs are a disaster, taking many years to reach maturity, requiring tight conditions for successful hatchings, and needing expensive housing with generous volumes of seawater for each adult creature. Only turtles and elephants—or humans—are less desirable if you want results in your own lifetime.

Marder's lab did make one foray into studying the developmental aspects of the stomatogastric ganglion, looking at the apprenticeship of the juvenile ganglion, which manages to run rhythms similar to those of the adult ten times its size. For this work, the lobsters were raised in the lab from eggs. Unless they have a client like Marder, with a research permit, fishermen are supposed to throw back any female lobster carrying eggs. The eggs, hundreds of thousands of them, develop slowly on the female's abdominal appendages, for as long as ten months before hatching in the case of the clawed lobsters. The larvae are transparent at first. In the open ocean, the larvae grow a bit, go down to the bottom to molt, go back up near the surface, grow some more, and repeat that cycle several more times. In a laboratory, they have to be kept for weeks in a special tank with constantly swirling water. In both environments, many of them just die. After a metamorphosis to an adult lobster form, the surviving inch-long juveniles look for niches at the bottom. Some five years and many molts later, they weigh about a pound, are sexually mature, and can be called adults. Painfully few of them make it. In the development work, of course, researchers weren't waiting for maturity, but even so, and even for the most talented graduate students, the experiments were difficult, and after a few years Marder called a halt.

The lobster may never be used as an ideal genetic model because of this long drawn-out life cycle and because, in the wild, eggs are usually fertilized before they are released, whereas transplanting genes requires access to unfertilized eggs. Although it used to be essential to breed animals to analyze a mutational effect in the succeeding generations, recent genetic tools can be used to target the genes of interest in the adult animal. Lobsters and crabs have surprisingly big genomes, twenty times that of the fruit fly. No one really knows why, and no complete analysis of those genomes is yet available, although by 2015, annotated transcriptomes for *Cancer borealis* and *Homarus americanus* were complete.

Today, researchers using crustaceans generally only study adults obtained from fishermen, commercial suppliers, or even from markets. They are kept near the lab in cold seawater tanks for only a few days or weeks. Unlike vertebrate animal models, where every mouse or rat is labeled with its genetic strain, date of birth, and other details, the crustaceans are whatever they happen to be, and the researcher just picks up the nearest specimen from the tank.

As long as you don't want to specialize in developmental or genetic neurobiology, there is a remarkable advantage to getting your animals from fishermen. Each lobster or crab, right up to the moment it was trapped, has been a successful animal, avoiding predators and functioning naturally and adequately. The same cannot be said for the hordes of lab-bred animals that are kept in unvarying and unchallenging conditions unlike anything in the wild. All of Marder's data come from animals that are well adapted to their natural environment, an important point, together with the fact that animals caught from the wild reflect wide genetic diversity, unlike the carefully bred strains of lab animals. Marder is therefore able to have confidence in the range of data her lab's experimental work has produced; whatever the phenotype of each individual animal, with perhaps slightly different neurophysiology or atypical neuroanatomy, it is evident that its stomatogastric ganglion worked adequately. As Marder points out, all these lobsters and crabs are comparable, in a sense, to the human population of individuals with widely disparate genetic makeup and developmental histories.

Judith Eisen, who was Eve's first doctoral student, told me they never knew how old their lobsters were. They were mature, of course, because fishermen have to respect minimum size standards, but lobsters grow throughout the decades of their lifespan, molting only once a year after they have reached a certain size. I asked Eisen whether they eliminated any specimens because they were too old. "When I was in Eve's lab, we mostly used the spiny lobster, flown in from the West Coast, and that's a warm water lobster so maybe they grow a bit faster than others. You don't know how old these lobsters are, but as long as they're still moving around and eating, they still have to have the circuitry functioning." Marder adds that they were unlikely to be past their prime. "We worked on lobsters weighing much less than the two to three pounders people eat, and some lobsters in aquaria grow to fifteen or twenty pounds."

A further influence on a research lab's choice of model animal has been the pressure to investigate aspects of biology as closely relevant to the human being as possible. "Translational" is the name of this game; it just means getting results that can be translated into ideas for understanding human biology, and especially for finding therapies for human disease. The lobster wouldn't seem to score highly by this yardstick, but research results described in later chapters are relevant to human health topics such as spinal injury or epileptic seizures.

When planning research, most biologists would nowadays give priority to a model that has been much studied by others. The amount of data stacked up on the physiology of the animal allows new investigations to start from a higher floor. Most people's work benefits enormously from having the answers to basic questions ready to hand. Nowadays that often means the model animal's genome has been sequenced and interpreted, but at the very least, the basic problems of how to manage the animal in the lab, how to prepare samples, and how to get around its weaknesses should have already been solved and a good body of "literature" should be available. On the downside, the most used preparations of the most used animals sometimes scarcely allow for new ideas. Despite the model's disadvantages, the crustacean stomatogastric ganglion has a long track record in neurobiology—and a distinguished one. Among the first researchers to turn to it was Sigmund Freud. Between 1876 and 1882, while he was a medical student at the University of Vienna, Freud used a new stain, methylene blue, to identify nerve fibers and study the structure of nerve cells in the crayfish stomatogastric ganglion. He worked long hours at the microscope, made detailed drawings, and published several papers.

After that, although there was a lot of research on crustaceans, the stomatogastric ganglion didn't attract much interest until the 1960s, when Don Maynard looked at the cardiac ganglion and then at the stomatogastric ganglion. Presciently, he identified one of the most salient points of interest, writing, "Perhaps the most conspicuous feature is the large number of functions which any single neuron may perform, i.e., act as 'pacemaker,' 'interneuron,' receptor, and motor neuron." For many years, researchers studying vertebrate nervous systems supposed this multitasking to indicate that invertebrates had necessarily to compensate for their small number of neurons by using each neuron in several ways. They took the daunting complexity of mammalian neuronal networks with colossal numbers of neurons to indicate that multifunctionality was unnecessary; there would be specialized neurons for each task. However, it is now accepted that both vertebrate and invertebrate animals use similar cellular mechanisms and circuit operations in their nervous systems.

For good reasons, biologists tend to specialize in one preparation, often throughout their career. Marder is not exceptional in this respect. Only if a research question that intrigues the scientist cannot be answered using that model will the leap into the unknown be made. It takes a strong streak of dissent—and a great deal of energy and time—to go out to look for a new, different, and more nearly perfect animal model.

Sticking to a specific model preparation implies accepting its limitations but gaining in depth. Of course, you need other labs and researchers to increase the sum of shared knowledge and to support and criticize your work from an expert point of view. About twelve labs are now working on the crustacean stomatogastric ganglion; compared with some fields in "consensus" biology, it's a small number. The labs hold a stomatogastric ganglion group meeting every year, the day before the annual Society for Neuroscience conference opens. It is a relaxed and friendly day of animated discussion and presentations of new work. Everyone knows everyone else; it's like a club. At least three generations of scientists are present, and most of the labs represented are led by scientists who trained or did their early research with the more senior scientists in the room. The openness of the discussions is striking because at meetings between labs working on similar problems the atmosphere is often cagey; they may be colleagues, even friends, but they are locked in competition for first discoveries and funds. They are often reluctant to share tips. This doesn't imply that the stomatogastric ganglion group is a band of angels, only that their research demands so much time-consuming and skilled work that it would hardly be worth rushing away from the meeting with someone else's idea and trying to run with it. People talk freely, unafraid of being scooped.

Many of the neuroscientists in the group attribute the collegial atmosphere at least in part to Marder's influence, her example of good scientific behavior, and her commitment to the principle of sharing scientific ideas. At a stomatogastric ganglion group meeting, the room feels pleasant and alive with a sort of exuberance exemplified by the rich variety of investigations being presented and the high spirits of friends who can't wait to catch up with each other. I suddenly realized that throughout the day, all the presentations and work being discussed were influenced by Marder's ideas: neuromodulation, the reconfiguring of circuits, homeostasis, and variability.

Whatever the limitations of working with crustaceans, when colleagues— or journalists—ask if she ever thinks of switching to a vertebrate model, the mouse perhaps, Marder is in the habit of replying that when she runs out of big unanswered questions that can be studied in the stomatogastric ganglion, and when mouse research has caught up with hers, then she'll think about it. She told a journalist, "the stomatogastric ganglion nervous system remains in my eyes the premier preparation for asking how circuit dynamics relevant for behavior depend on the properties of the cells and their synaptic connections."

4 Marking Time

Doctor Marder started the year 1975 in the forests of Oregon. She had a postdoctoral fellowship from the National Institutes of Health for work with David Barker at the University of Oregon in Eugene. I asked Brian Mulloney whether her choice had surprised him. "Not at all," he answered. "Barker was a young successful assistant professor in a very strong department with a good tradition." The word "tradition" principally referred to Don Maynard, who had introduced the crustacean stomatogastric ganglion to the world of neurobiology, and Graham Hoyle, one of the major figures in insect neurobiology and small systems. Sadly, the department was in the shadow of Maynard's sudden death two years earlier. His team had been redistributed among his colleagues, who included his widow, Teddy Maynard, an accomplished microscopist.

Mulloney went on: "Barker had a good group and was very amiable. And he was in tune with Eve's line of thinking." Barker's lab was working on the inputs to the stomatogastric ganglion descending from the esophageal and commissural ganglia in the stomatogastric nerve. Marder thought this work might provide the key to understanding some of her own experimental results. Although she had spent her time in Selverston's lab identifying neurotransmitters in specific neurons, part of her mind was always occupied with the unexplained existence of the many active substances reaching the ganglion that set off unexpected neuronal responses, like secret agents. When Barker had visited the neurobiology labs in San Diego, Marder had been surprised and encouraged by his recognition of what her results implied and by his quick grasp of what she was fumbling toward.

Particularly mystifying were substances such as dopamine or serotonin, which were among the ones Marder had tried dumping on the ganglion. There was no trace of them in the neurons of the ganglion, so they seemed unlikely to be used as neurotransmitters there. Barker's work showed that the stomatogastric nerve, which runs into the middle of the neuropil, contained precisely those molecules: dopamine and serotonin, as well as octopamine. Today, dopamine and serotonin are almost household names, but octopamine still doesn't get much attention. Its name is a giveaway: it was discovered in octopus salivary glands in the late 1940s. In the early 1970s, Barker had found octopamine in the lobster when he was working as a postdoc with Edward Kravitz at Harvard. When Marder showed him the puzzling results of her experiments and told him that her data didn't make sense in terms of neurotransmitters in the ganglion, he immediately realized that her results reflected a true physiological phenomenon; in the living animal, those substances would be brought to the ganglion by the incoming nerve or as hormones in the hemolymph. It was not just an inexplicable experimental finding; it was a significant clue. Like Marder, he saw the as yet unexplored possibility of chemical modulation acting alongside the direct neurotransmitter messages across synapses.

Marder intended her job in Eugene to be temporary, a holding position for a year while her boyfriend finished his PhD with Selverston in San Diego. They both wanted to go to France, where there were several excellent laboratories for invertebrate neuroscience. Neither of them was happy in the U.S. campus atmosphere of bitter divisions over the Vietnam War.

She started off in the lab keen as mustard and with the brisk no-nonsense approach of someone who has little time to waste:

> Jan 8 1975 I am planning a series of experiments with D Barker to demonstrate that acetylcholine is released from the dorsal dilator muscle. [The dorsal dilator muscle is the target of the pyloric dilator neuron she had studied.] This would satisfy one of the two remaining criteria left untouched by my PhD thesis. Most importantly, if this technique works, it will be possible to quickly screen the rest of the stomach muscles for acetylcholine release. This would be very nice because then it would be possible to make stronger statements about the electrically coupled neurons using the same transmitter.
>
> Eventually, the experiments I want to do are as follows:
>
> 1. Demonstrate release of ACh from the d.d. nerve/muscle.
> 2. Use LP muscle as control—nothing there? [The lateral pyloric muscle is the target of the lateral pyloric neuron. The LP did not contain choline acetyltransferase so Marder did not expect its target to release ACh.]
> 3. Screen all the stomach muscles.
> 4. Publish.

To tidy up the unfinished business of her work, completing the requirements of her little bible of proofs, Marder wanted to prove that the neurons she had identified as using acetylcholine actually contained it. Although she had shown that the enzyme that made acetylcholine was found in PD neurons, she had not demonstrated the presence of acetylcholine itself. Conversely, she had never found any choline acetyltransferase activity in the LP cell, so if she could show there was no acetylcholine to be found in it, nor in the muscle it controlled, then the LP certainly used some other neurotransmitter (she suspected it was glutamate). That too required more work, and she was looking for techniques to do it. So another reason for going to Eugene was that Barker practiced the "hot zap." In fact he was one of the developers of this way of identifying neurotransmitters. How did it work? First, the saline solution bathing the ganglion was loaded with radioactive precursor molecules for a particular neurotransmitter. Neurons using that transmitter took up precursor molecules from the bath and used them to synthesize transmitter molecules—radioactive transmitter molecules. Then molecules of the candidate neurotransmitter were separated out by high-voltage electrophoresis. If they were radioactive, it meant they had been synthesized inside the cells of the ganglion during the incubation with the radioactive precursor molecules. It sounds neat, but the high-voltage equipment was big and bulky and risky in use; within a few years, the hot zap was abandoned in favor of tamer techniques.

Marder found it difficult to pin Barker down to a program of work with her. The lab books from Eugene are full of seemingly unconnected experiments, some proving fruitless: "not a rip-roaring success" as she noted of one of them. Marder was using "iontophoresis," in which a microelectrode filled with a candidate neurotransmitter accurately applies the substance onto identified neurons. She was trying to differentiate between the effects of histamine, acetylcholine, dopamine, serotonin, GABA, and glutamate so that she could describe the distinguishing features of the responses. She was amassing data that she could not yet put together; there was no new conceptual breakthrough that year. "I don't think my ideas moved forward … I was still collecting. I was fascinated by what the different transmitters in different cells might be … I just figured I had to do iontophoresis on the cell bodies, and I figured out that the same cell had receptors for tons of different things. I saw that very clearly, but there was nothing publishable. Bits and pieces."

Away from the bench, she wrote a long, thoughtful report on her San Diego research that was published in 1976 in *The Journal of Physiology*. In the months since tying up her thesis, she had gained a clearer perspective on the work. Although the thesis focused on the PD motorneuron and its dorsal dilator muscle, Marder had plenty of evidence, if less complete, for the other motorneuron-to-muscle pairings. In both the *Nature* report and her thesis, she had wisely stuck to her "nice little story," which recounted the most successful part of her work. Presenting the rest required more thought and analysis, and she tackled it in Eugene. Her statement of her goal in this paper was particularly cogent:

> The functional significance of the diversity among neurotransmitters is not understood. Why many different neurotransmitters are used in nervous systems and why any one neurotransmitter is used by any given neuron are unanswered questions. One approach to these questions is to characterize the neurotransmitters and receptors used by all the neurons within a physiologically complete nervous system. This would determine if there are any correlations between the functional relationships of the neurons within the nervous system and the neurotransmitters used by those neurons. This paper reports the first results of an attempt to characterize completely the neurotransmitters within the stomatogastric system of the lobster, *Panulirus interruptus*.

In the paper, Marder presented her data showing which of the stomatogastric motorneurons seemed likely to use acetylcholine to excite their muscle targets and which probably used L-glutamate. She pointed out that there were also electrical connections in the ganglion and discussed whether neurons that are electrically coupled might be constrained to use the same neurotransmitter, an idea Brian Mulloney had first suggested. She examined the question of why stomatogastric system muscles should be excited by acetylcholine while other crustacean muscles that had been studied, in claws for instance, were activated by L-glutamate; could this be related to the unusual (for an invertebrate) lack of inhibitory input to the muscles? She ended with the question of whether the motorneurons she had shown to use acetylcholine to excite muscle would use that same neurotransmitter to make their very different inhibitory connections in the neuropil. The question was posed without emphasis, but obviously, if a different neurotransmitter were used, that would mean a neuron could use two neurotransmitters. And that would constitute a definitive flouting of Dale's Law with a bright feather in the cap of the investigator who could show it.

* * *

Marder liked living in Eugene. In her first experience as a postdoc, she had a new position in the team, now expected to help students and take more responsibility in the lab. She wrote to Mulloney, "In the last few weeks I've begun to understand how much I owe to you for all the advice and encouragement I got, especially in the beginning. I see it now because I see myself flipping roles, and there are a bunch of students here, and I realize how unbelievably patient you were with me. My father's mother once told him that children never repay their parents' love, but they in turn are parents to their ungrateful children, so I guess to a certain extent goes the science heritage—anyway, over a little time and distance, I can more clearly see how much I do owe you, and thought I'd say it once, for you."

Marder netted herself a Helen Hay Whitney fellowship that allowed her to work anywhere in the world without having to ask the host lab for a salary. She wanted to work in Paris, at the Ecole Normale Supérieure, with the American scientist JacSue Kehoe. Marder enormously admired Kehoe's research, saying her work in the early 1970s was "the absolutely most beautiful invertebrate pharmacology anyone had ever done." Kehoe's published papers showed that at least three different types of acetylcholine receptor could be found on the same neuron, and that their responses to the neurotransmitter followed different time courses. The luminaries of the Laboratoire de Neurobiologie included its director, Hersch Gerschenfeld, Anne Feltz, and Kehoe's husband, Philippe Ascher. Marder was eager to learn from them: "At that time, if you wanted to do invertebrate neuro-transmitter physiology and pharmacology, that group was clearly the best in the world on objective grounds."

Marder adds that, just as important, she wanted to observe an independent woman scientist at close quarters; she wondered how Kehoe could do such elegant work and also have a family—she must have written the papers while coping with a new baby and a toddler. At twenty-seven years old and despite her undoubted passionate interest in her science, Marder was trying to picture her future, especially her personal life. As she now says, at the time you could hardly fail to notice that the older generation of women scientists had almost entirely eschewed family life and that the generation before that had been frankly odd, pushed to the margins as bluestockings and scarily dedicated to their work.

From San Diego, Marder had written to JacSue Kehoe, and in 1973 they met at Cold Spring Harbor, where Kehoe was teaching at one of the summer courses. Marder was gratified when Kehoe invited her to collaborate in writing for the *Annual Review of Pharmacology and Toxicology*. A review is a compilation and critique of as much of the research on their subject as the writers can handle. It is also something of a think-piece, and in this case it allows us to see what Marder knew at this precise point in the story. It was titled "Identification and Effects of Neural Transmitters in Invertebrates" and dealt with sea slugs and snails (gastropods) and insects and crustaceans (arthropods). Marder trawled thoroughly through the recent publications, and 111 papers are referenced at the end of the review. That looks like a lot of work, and it was. However, it would be dwarfed by the number of references you would find today, and that highlights one of the effects of the information explosion on young scientists, an effect Marder deplores without knowing how to help them.

> Partially it's because the landscape is so much larger, but also they don't have good ways of navigating it. It wasn't difficult for me to read everything that was known about invertebrate transmitters, at least what I thought was relevant. How would a student today know everything that was relevant to what he or she was doing? The way I did searches in the old days, we all did, you went to the library, you got a stack of journals, and you read the tables of contents. And so if you hadn't gone to the library for three months, you sat and read all the tables of contents in all your journals for those months. In this way, you developed a sense of what was known, where it was likely to be, how to find it—and you accidentally came across things that you might not have known were there. I watch my graduate students and postdocs look for things and not actually know what's out there, what's relevant, because they're doing targeted internet searches rather than reading journals.

Marder made a long table of the different invertebrates and what was known of their neurotransmitters. One of the interesting things that stands out from this table is that researchers were seeing many neurons that contained several neurotransmitters. The candidate, or presumed, neurotransmitter was usually present in much larger quantities than the other substances, but was their presence a clue to unexplained complex mechanisms or a result of experimental clumsiness? A paragraph from the review conveys the flavor of her critical approach:

> Virtually all of this work has been done on hand-dissected neurons, which certainly contain an unknown amount of glial and likely even neuronal contamination. With increasingly sensitive assays, the small amounts of substances contributed by contaminating tissue become more significant. In injection studies, leakage could be a problem. Second, even if low levels of substances are actually present in the neurons involved, are they physiologically significant? Phrased differently, this question asks, how useful are biochemical techniques alone for transmitter identification?

Here Marder was emphasizing that it was an important part of transmitter identification to verify that, as well as being found in a neuron, a substance had a function. Electrophysiological techniques could measure a follower neuron's reaction to that substance, something biochemical techniques obviously couldn't do. If the substance elicited a reaction, then it was probable that it had a function in the system.

In reference to the work of Hanley et al., the transmitter function of 5-HT (serotonin) has been confirmed by physiological studies whereas the acetylcholine reported in this cell has not yet been shown to be utilized in a monosynaptic connection made by this neuron. These data point out the caution that must be exercised when drawing conclusions from biochemical data alone, and stress the importance of combining physiological and biochemical techniques for transmitter identification.

I don't know whether that paragraph was written by Marder or Kehoe, and for once, Marder couldn't help me. However, her lab books show she had already thought about Hanley's work on snail neurons, and this kind of scrutiny of results with a keen awareness of what may be misleading has become a recognizable feature of Marder's writing. Only by critical thinking about possible sources of error over the years could she have realized decades later that some measurements, rejected as errors, were not errors at all but important clues indicating a wider range of physiologically valid mechanisms than anyone had expected.

The review also reflects the mid-1970s: a time when neuroscientists first recognized the unexplained variety of neurotransmitters and the wealth of receptors and receptor responses that they would have to investigate. For example, the neurotransmitters "firmly established" in molluscan neurons were acetylcholine, serotonin, and dopamine, whereas in arthropods they were acetylcholine, GABA, and glutamate—but there was no functional explanation of the difference. Marder was beginning to make a real intellectual contribution to this significant work. She was intrigued by the difference between a neuron's response to molecules of neurotransmitter transferred across a synapse—a direct message—and its response to other substances binding to receptors elsewhere on its membrane. This kind of response might alter the neuron's electrical properties even if it didn't directly trigger action potentials in it. The evidence that David Barker's lab was finding showed that in various invertebrates, including lobster and crab, octopamine was secreted by neurons into the "environment" of other neurons. Researchers had observed that octopamine affected neurons in several invertebrate preparations. In the final sentence of the review, Marder explicitly used the term "neuromodulator" for the first time in print: "All of these data are consistent with a neuromodulator role for octopamine in these systems."

The review was sent off in November 1975, and Marder left for Paris in December with a couple of suitcases and a supply of her trusty brown notebooks. I asked Mulloney what he had thought of her choice this time, and he answered, "Paris was not a surprise, obvious thing to do. Going to work with JacSue was a fabulous opportunity." He added that he had even been envious.

* * *

Arriving at Charles de Gaulle airport, Marder was dismayed—vexed—to find that she couldn't understand the public address announcements. Fortunately, JacSue Kehoe was there to pick her up and shepherd her into Paris. The next morning, she went to the Laboratoire de Neurobiologie of the prestigious Ecole Normale Supérieure. In the lab, everyone spoke French, except when speaking directly to Marder. For a year, she says, it was a daily battle; unable to fully understand the conversations around her, she struggled to grasp the science underlying her colleagues' work.

In fact, she hadn't given the French language a thought for years. I suspect that she had learned it as a word game, much as she had learned Hebrew, and I recognize the pattern because that's exactly how we learned Latin at my school in England—not as a language: the analogy is more with building blocks or a code. Marder had learned French in a literary way at school and college. Her written French had been excellent, but now she was appalled to find she couldn't make out spoken words that she would have instantly recognized on paper. She had a strange sensation that the French she had learned at school was in one part of her brain, to which she didn't have access, while she struggled with spoken French in a different part.

There was another reason that she didn't understand much and that, too, was a bit of a shock: she wasn't prepared for what the Paris lab would actually be doing. On arrival, she realized it was a biophysics lab, and the research involved a lot of quantitative biophysics, noise analysis, and kinetics of ion channels, all of which was outside Marder's experience. As she says, "I wanted to learn how to do invertebrate neuropharmacology from them. They were all doing biophysics, and they had become interested in second messenger pathways and signal transduction—early, before it was called that. All my formal university coursework was in biochemistry and molecular biology. I had become a competent electrophysiologist, technically. And I was even a competent *cellular* electrophysiologist. I did not know any biophysics."

In that era, Marder says, biophysics attracted people who tended to be smarter than the biologists who were drawn to more descriptive, fuzzier, less quantitative work. That led to a certain elitist arrogance on the part of biophysicists because it was so hard to know anything at all, then, that quantitatively satisfying or rigorous work was the exception.

It was a real divide. I can't do better than quote a Nobel prizewinner's speech here. In 2000, Paul Greengard shared the prize for work in the 1970s revealing the mechanisms by which receptors pass signals into the cell. He said,

> At the time that we started this work, neuroscience was not a clearly defined field. There were two types of people studying the brain. There were biophysicists, working in physiology departments, who believed that everything significant about the brain could be explained in terms of electrical signaling. And there were biochemists working in biochemistry departments who would happily throw a brain into a homogenizer, with as much abandon as they would a liver, and look for enzymes or lipids. But these biochemists were rarely interested in brain function. And so these two groups rarely spoke to each other, which is just as well because when they did they didn't have nice things to say.

Unfortunately, Marder had come up against this divide, and it complicated her work in Paris. Her superiors were young (young for scientists—in their late thirties and early forties), and they had rebelled against the traditional structures of French science to start a new lab dedicated to quality cellular neurobiology and biophysics. They were opinionated, of course, and they were fine scientists, as well as friendly, cultured, and talented conversationalists. But they were not entirely receptive to the research questions Marder was asking. Essentially, with great technical skill and exacting intellectual analysis, they were investigating the activity of the structures underlying the neuron's communications: its receptors, ion channels, and synapses. Marder was always focused on the communications between the different types of neurons in the pyloric circuit, the functions of each neuron, and how those functions were integrated in the network. She wanted to identify neurotransmitters and describe their effect, how the receptors changed the receiving muscle or neuron's properties, and then to understand how that affected the activity of the circuit. Unlike her Paris colleagues, she could not be satisfied with understanding the fine detail of the assemblies of molecules forming receptors and the ion channels they influenced. For her, discovering the molecular structure and cellular properties of an individual synapse would only be a means to the end of understanding a broader picture of how circuits worked.

One thing Marder had thought about and checked before traveling was the equipment she wanted. She found out that the Paris lab had no amplifiers for extracellular work, so she had a set made by the technicians in San Diego. With these amplifiers, she could hope to recognize the signature tunes, the characteristic firing patterns of neurons whose axons were in a motor nerve, picked up by a wire electrode placed close to it. She had to have them to identify the neuron she wanted to investigate before piercing it with an intracellular microelectrode.

Alas, when she got to Paris, she soon discovered that her special amplifiers were substandard, with a lot of noise in the signal. That was far from her only problem: the lab's equipment was appropriate for work on the Californian sea slug, *Aplysia*, and the intracellular amplifiers and stimulators had no labels on the knobs and no instruction manuals. She didn't speak enough French to ask for help. With the dissecting tools she'd brought from the United States at the ready, she sat and stared at the stuff and at the unfamiliar crab species she had just bought in the market.

The Paris markets delighted her, but the great range of gastronomically desirable shellfish on display was mostly unfamiliar. Marder explored the fish stalls in the famous market on the rue Mouffetard, with instructions to find the cheapest suitable crustacean for her work. In the tanks, she saw spiny lobsters from Mediterranean and South African waters, relatives of *Panulirus interruptus*, but so staggeringly expensive she wondered how anyone could afford to eat them, let alone use them as research animals. She tried crabs, crayfish, and occasionally the American lobster's European cousin, which lives in the North Atlantic off the European coasts. January 1976 went by on these trials. The stomatogastric ganglion preparation from the Atlantic lobster, *Homarus gammarus*, is different from the spiny lobster she was used to, and an echo of homesickness crept into her lab book: "The STG has a very fine sheath, which I removed, but sits in a cylinder over a section of nerve, is hard to illuminate and hard to pin and not anywhere as pretty as the STG in *Panulirus*." Marder settled on working with crabs.

It was a flaky start. Moreover, within a month, a French woman carried off her boyfriend. Marder was relieved—the relationship had been complicated—but she found herself, as at earlier times in high school, feeling isolated. She wrote long letters about her experimental tribulations to Mulloney.

I am trying to follow the development of Marder's thinking, but in Paris, it seems to be masked by the practical difficulties, the obstacle course she faced. It is almost comical as she recounts it, but at the same time my heart bleeds for the young Marder. She immediately hated the lab's manipulators: "these workhorse things that were just beneath contempt … they were mounted square and so the electrode had to go straight down and so then the researchers, to be able to see with the electrode going straight down, they had the microscope tipped. That's okay for the *Aplysia* ganglion because it's big with very big cells that are orange pigmented and so they had enough contrast to see the electrode through the tipped up microscope." The crab stomatogastric ganglion is flat and translucent, so this setup made it difficult to see what she was doing, especially as the electrodes were much bigger than the ones she was used to.

These electrodes were made ingeniously and skillfully, but not at all the way she was used to; when Marder first saw the process, her mouth dropped open. The starting material was a thick-walled glass tube almost as thick as a pencil. The tube would be heated over a Bunsen burner and pulled apart. One half was clamped vertically in a mount, and an angled microscope was positioned to control the rest of the procedure. A heating filament was brought into contact with the pulled tip of the glass tube, and the tip was bent to form a little hook. A small weight made of solder was hung on the hook, and then, still watching through the microscope, the researcher used a manipulator to bring a heating coil up to the tube, which began to soften. The weight pulled it down, elongated it, and eventually it broke off. The weights could be chosen to produce the kind of tip required. It was laborious, it was a craft, and it took months to become proficient. Kehoe was a virtuoso and could make any electrode of any shape or tip size in a few nonchalant minutes. Marder struggled. On the plus side, a successfully finished electrode had the great advantage of having thin walls only at the very tip, with no extended area of narrow shank like the fine-walled electrodes she had made before. The electrode thus had excellent electrical properties. After extolling the virtues of these fastidiously crafted electrodes to me for a few minutes, Marder breaks into a doleful laugh and says, "The problem is, they're physically humungous. They get in the way so you have visibility issues trying to see around them."

To cap it all off, the paper film she had always used to record traces wasn't available in France. The paper film could tolerate red light; now she had to develop film in absolute darkness. Her Paris lab books bulge with the celluloid strips, taped or stapled down, instead of the paper traces pasted in neatly and on which she could draw arrows or write notes.

It took her many weeks to master the paraphernalia. When anything went wrong, she had no idea whether it was her mistake or the equipment. Were her electrodes killing the cells, was it the manipulators that weren't fine enough for her preparation, or was she being clumsy? Often she couldn't see the preparation well enough to be sure. She seemed to have lost her competence, lost her easy mastery of experimental technique: an unsettling feeling.

Outside the lab, Paris enchanted her. Kehoe had found her a "chambre de bonne," a top-floor maid's room, the garret of every student's dreams, on the beautiful Ile St-Louis in the middle of the Seine. In the morning, Marder walked over the bridge, with Notre Dame on her right, and up the hill to the lab. Her window looked out on the rooftops of Paris. Inside, she had about 200 square feet, which was enough for a bed, a table, and a chair. There was a sort of alcove with a hotplate and a minuscule fridge. Quite rare for a maid's room at the time, she had her own tiny *salle d'eau*, where the shower had a drain in the floor right next to the wc. It had the merit of being private, but every shower led to a flood. A more serious inconvenience was that there was neither a phone nor a doorbell, and it was a fifth-floor walk-up. The only way someone outside could reach her was to ring for the concierge or wait for someone going in who would take up a message.

One day in late summer 1976, she came home to find the lock broken, her camera and some cash stolen, and instantly the bohemian charm evaporated. Luckily, another apartment that Kehoe knew of was available, and Marder moved even closer to the lab. Rue Descartes was a wonderful location, and this time she had a big ground-floor room with a real kitchen and bathroom – but still no phone. However, the notoriously long waits for telephone lines would soon be over. I was also living in Paris at the time, so I know that, by order of President Giscard d'Estaing, the streets were in chaos with the trenching works for cabling to bring Paris up to date *tout de suite*. A few days after moving, Marder walked out of the building and saw a phone installer at work, so she asked him for a line. "*Tout de suite*," she said. A week later, she had her phone.

Marder quickly realized that Kehoe had far more help in bringing up her children than most American scientists could hope for. There was a nursery school on the way from her home to the lab. She had a housekeeper to help look after the children. However, a more important factor in Kehoe's comfortable combination of family life and research was that in the French system of dedicated research institutions, she had no teaching, no administrative responsibilities, no students or lab team. Marder was her first postdoc and her last. None of this would be attainable in the U.S. system, so the question of how to be a woman scientist with a family remained unanswered and disturbing. In July 1976, Marder wrote to Mulloney about her fears of becoming "an honest to god hot-shot" but alone and unloved. Marder was, and still is, always sociable; she was used to a circle of close confidants, with larger circles of friends and colleagues rippling out. Here she had fewer friends, and that first summer she was unprepared for the great August exodus of Parisians. To Marder's surprise, almost everybody went away, and Paris was virtually closed up. Even the maker of the most sought after ice sorbets, la Maison Berthillon, famed throughout Europe, perversely closed in August. Marder hadn't planned a summer vacation so soon after joining the lab, but that was obviously a mistake.

During this lonely month, her problems in the lab compounded her personal unhappiness. It was an intensely testing time. After this, I think, she would avoid ever putting herself in such a vulnerable position again. She wrote to Mulloney in August, a letter that reflects her resolve to pull herself out of trouble and carry on her science: "I went through a very bad crisis, decided I couldn't work without decent equipment and *Panulirus*, and almost called you to ask you if I could come work for 2 months. It's absolutely frustrating. With 10 *Panulirus* and a real equivalent to what I had in San Diego, with what I know now I could do some really simple good experiments. Well I'm not as discouraged as I was, but it's a drag. Have been having trouble with the crabs—that is they don't work much … there is virtually no spontaneous activity."

Her problems at the bench didn't go away, but September came at last, and her friends came back. A fellow postdoc, working with Anne Feltz, was an Englishman with a Belgian wife, two children, a remarkable sense of humor, and the ability to speak a sort of fluent Anglo-French. Paul Adams was brilliant, Marder says, a fine pharmacologist, and she talked to him a lot. Their conversations were often cut short when he had to catch his train home to dinner in the suburbs. He was famously absent-minded, and, to protect the family purse, his wife doled out a little money to him every morning. If she was away, he would lose his sense of time, work all hours, stay nights in the lab, and run out of cash or forget where his wallet was. His lab mates used to feed him in the many small restaurants of the Rue Descartes, and the conversations went on uninterrupted.

The terms of Marder's Hay Whitney fellowship required her to report back every year, and so in autumn 1976, Marder went back to the United States. She took the opportunity to go to the Society for Neuroscience conference. She also went to interviews for assistant professor positions at some East Coast universities. By today's standards, this would have been premature, just two years out from her doctorate and with a grand total of three published papers. But then, for her generation of young neuroscientists, the doors were wide open as universities responded to the increasing prominence of neuroscience. I think she was also motivated by her frustrations in Paris and impatient to have her own lab because she had discovered that she didn't like working in financial dependence on other people. In 1977, there are more signs of restlessness: four six-week gaps between experiments logged in her lab books that she explained to me as due to interviewing for jobs, going to a meeting in Italy, traveling for three weeks with a recently widowed aunt, going to the Society for Neuroscience conference, and reporting on her fellowship. All the same, it adds up conspicuously to half the year going by without significant work at the bench.

Marder was sharing an office and a precious telephone with Paul Adams. He was also looking for a job in the United States, and every week they went through the notices at the back of the journal *Science* and sent off their applications. Adams was hoping to find a job somewhere like Galveston because his wife yearned for a warm climate. A lot of time seemed to go by with no offers. One day, the phone rang, and Adams had a brief conversation. He rang off saying, "Hey Evie, I got this job offer—Cornell!" Twenty minutes later, the phone rang again, this time for Marder. It was the chair of the department at Cornell, telling her she was their second choice, so she should not accept anything else: "We really like you." He hadn't realized he was dialing the same phone number twice. They laughed, but they were relieved, although in the end neither of them took that post.

Soon Marder had two job offers, the one from Cornell and one from Brandeis. On May 5, 1977, she wrote a letter from Paris: "Dear Brian, Well what can I say except that I took the job at Brandeis... needless to say that I am very happy and feel that I made a good decision, and I won't have to start until Sept 1978, so all is taken care of."

Her choice betrays a certain discomfort with the formality of France. She imagined herself, a new assistant professor, wearing jeans, sitting on the floor in a corridor talking to students, and knew that no one at Brandeis would even blink. But that wouldn't go over so well if the chairman of the department at Cornell walked by in his tie and jacket. Also, Marder wanted independence, wanted to be free of interference and have the luxury of making her own mistakes in private, so it weighed in the balance that there were no "real neuroscientists" as she called them at Brandeis yet. Cornell already had a flourishing Neurobiology and Behavior Department; more people would have been looking over her shoulder.

The Cornell job had offered $19,000, whereas the Brandeis salary was only $16,000. Marder didn't want to go to Ithaca, and she didn't intend to make this big decision on the basis of salaries, but she did call the chair of the Biology Department at Brandeis to tell him Cornell had proposed more. He said she could take the offer or leave it. Marder's meagre experience had not prepared her to handle a salary negotiation; she had grown used to France, where there was a national pay scale that nobody questioned. She had no idea that salaries at some universities might be negotiable and that she might have been able to get more. The incident still rankles because, as she says, "The sad thing is, the years that you need the money are when you're young … when I had a mortgage and a second mortgage on my apartment and at the end of the month I was counting the pennies to put in the toll booth."

* * *

At the beginning of her second year in Paris, Marder made friends with a young woman in Gerschenfeld's team, Danièle Paupardin-Tritsch, whose PhD thesis on serotonin receptors in snails and *Aplysia* Marder had read and admired. Marder was anxious to work in better conditions on the experiments she had in mind, and Paupardin-Tritsch was looking for something new, so in January 1977 they started to work together. Paupardin-Tritsch was set up in a lab across the hall from Marder's bench space. Although they got their superiors' agreement to the collaboration, for some reason, they were told not to merge their equipment. Paupardin-Tritsch had fine manipulators and a good voltage clamp and perfusion system—a fine working rig in fact—but no extracellular amplifiers that would help identify cells. Marder found herself caught between two imperfect options. She could work on a nice rig with a smart and congenial colleague and get decent recordings of unidentified cells, or she could struggle on alone, getting less reliable recordings but of identified cells. Neither Kehoe nor Gerschenfeld saw it as a problem at the time. They were working on synaptic transmission, a mechanism then thought to be the same in all neurons, and so they could treat all neurons as equal and any neuron would do for their experiments. None of their work was related to the neuron's function in a circuit. Besides, they were working on *Aplysia* on cells that could be identified under the microscope just by their markings and position; they had no need for special amplifiers. They thought Marder could do the experiments she had planned with Paupardin-Tritsch and then repeat them on identified cells later if she thought it necessary.

But it was a real problem. Marder knew that the whole point of working on identified cells was that, if she knew which neuron had which properties and where it fit into the circuit, the results of her experiments would provide clues to network function. If she didn't know what cell she was working on, then she could learn only that at least this one cell had the response she was recording; she could "collect" the repertoire of responses but couldn't learn anything specific about the organizational principles of the network. The unsatisfactory equipment wasn't her only concern, however. She was also hampered by the preparation they were working on.

Marder didn't fully appreciate this issue until years later. Paupardin-Tritsch was pregnant and working short hours, so they usually didn't dissect out the whole of the crab stomatogastric nervous system with the three anterior ganglia; they just cut the stomatogastric nerve and the motor nerves, and took out the ganglion. This would have been perfectly fine with the spiny lobster as the ganglion would have continued its rhythmic action for a while. But in *Cancer pagurus*, the ganglion was usually silent without its descending input connections. In some ways that was helpful. Precisely because there was no spontaneous activity in the preparation, any activity they recorded must have been elicited by the substances they were putting onto it. And if they recorded activity, that meant that, somewhere, the ganglion had receptors for that substance. However, before they applied anything, they should have been able to record the distinctive firing rhythms of the neurons they were working on. The silence and the lack of any signature tunes made it all the more impossible to name the cells.

Nevertheless, they extracted a paper from this work, and Marder wrote to Mulloney in late November 1977, "Here is a manuscript which we sent to *The Journal of Physiology*. It was hard to write because we were forced to choose figures for the pharmacology not for the stomatogastric system, and I didn't want to confuse things with the details of stomatogastric stuff."

It was hard to write, but Marder knew she had to publish. She had accepted the job offer from Brandeis in May, and even in those more relaxed days, she couldn't afford too big a gap in her publishing record. With results in unidentified neurons, Marder could only present the work as descriptive of the phenomena observed, divorced from any consideration of circuit function, and the report is uncharacteristically woolly in this respect. Only one of the twelve figures refers to named cells, and only the last three paragraphs of the discussion, headed "Relevance to the Stomatogastric System," tie the results back to the cells of that system. The paper reads as a fiendishly complicated piece of forensic pharmacology and contains a more sophisticated analysis than her first studies. With the idea of multiple receptor types for each transmitter in mind, Marder and Paupardin-Tritsch tracked down the precise conditions in which cells responded to each of the three substances: acetylcholine, GABA, and glutamate. The paper was much cited, which is generally considered a sign of success, because it contained so many bits of useable information for other researchers. In particular, it hinted to the attentive reader that substances that bound to a particular one of the three types of acetylcholine receptor would turn on bursts of action potentials and activate the central pattern generator (CPG) in all crustaceans. But Marder couldn't state it explicitly "because it was unanchored in the network, it was just sort of indirect observation."

* * *

In her postgraduate and postdoc years, Marder had found fellowships to support herself quite easily. At Brandeis, she would have a salary as an assistant professor with teaching obligations, but she also had to think about supporting research in her own lab. She applied to the National Science Foundation (NSF) and the National Institutes of Health (NIH), the two biggest funders of biology research in the United States. Marder posted her grant applications in February 1978, memorable for a famous blizzard that locked in the East Coast. As the funders' deadlines approached, she knew her envelopes from France were somewhere out there in the mountains of backed up mail. She chewed her fingernails. Both applications arrived on time. The two applications were identical, and both institutions offered funds but with different conditions and dates for acceptance. Unfortunately, she got caught between them as the NSF program director made her an offer and forced her to make a decision before the NIH had responded. So she accepted the NSF grant for less money than she needed and later had to turn down a much higher grant from the NIH. That experience taught her another lesson, and she has always helped her departing postdocs in the funding jungle.

It seems that her time in the Paris lab was spent, as in Eugene, just "collecting," not advancing, not forging a new path through the forest. I can't find any thread of conjecture, any note of theoretical enquiry; there are no exciting tracks to follow through the pages of her lab books. Hoping she would give me a lead, I asked her whether, when she was writing her grant application, it was in a consistent continuum with ideas that had matured in Paris. Not really, she answered, she just wrote a research plan she thought was interesting. "It was actually about a very cool problem: the question of whether two electrically coupled cells might use different transmitters. And if they did use different transmitters, whether you'd get sharing of transmitters across the electrical synapses, whether the transmitter molecules would move or did the neurons have to keep their transmitter segregated. So it was really transmitter identification again. I was just curious about whether electrically coupled cells had to use the same transmitters and if not how they kept track of what was on which side of the channel."

So, no, it was not related to recent work; the subject had interested her before she came to Paris. Among the methods she proposed to use, to look for movement of radioactive choline across the gap junctions, was the hot zap. In some ways, she could have written that proposal in Eugene. Had she really been standing still in Paris? The Laboratoire de Neurobiologie was certainly an intellectually stimulating place, and Marder learned a great deal there, especially the membrane biophysics and pharmacology that have informed all her subsequent work. But it was a digression away from her cherished and entirely valid objective of studying circuit function, and the benefits of her time there only became evident gradually over the following years. None of it helped her develop her own ideas while she was there. She certainly felt, and still feels, that the experience gave her a more sophisticated perspective on the world, and she has recently written an article recommending young scientists to try working in other countries, at least for a short spell, arguing against the commonly held view that leaving the country damages your prospects on the career ladder. "In retrospect I realized that I received those [job] offers precisely because I had gone to France. The work I did there was not spectacular, but somehow the fact that I had been adventuresome enough convinced others I was ready to be independent."

Marder's dissatisfaction with her postdoctoral work caused her some apprehension at the prospect of having her own lab. She found it reassuring that her salary was actually for teaching as she knew she could teach; her experiences as a teaching assistant for several semesters in graduate school had been successful.

* * *

In early 1978, she and Danièle Paupardin-Tritsch turned their attention to the transmitter receptors on the muscle cells of the stomach and pylorus. Working only on the muscles, with no neuron identification problems, calmed her. "Boring but useful," she told me. Snatches of levity returned to the lab books: "From here on cell clamped, potential could no longer be changed because electrode hated me." "French glue is lousy." Two more papers would come out of this work, which Marder finished in Brandeis the following year, but it was not a glorious result for a period in her career when she should have been very productive.

By the time she left Paris, in October 1978, Marder could rattle away in French quite fluently—but incorrectly. She spoke as quickly as possible to blur the fact that she had no idea which words were masculine or feminine, and her French friends all thought she spoke French well. A fraud, she says. But she felt frustrated because she could hear the difference between her accent and theirs without being able to fix it. Too old, she now says.

She had been through three stages of adaptation. In the first year, she was often confused, not understanding enough of the language to grasp the culture. Then in her second year, she understood enough French to recognize that she was completely bewildered by the culture. For instance, her dentist scolded her for not ironing her jeans, and the secretary at the lab scolded her for wearing jeans with holes in them. Marder, of the Woodstock generation, was accustomed to wearing jeans until the legs had so many holes and were so frayed that you ripped them off at the knee and wore them as shorts all summer. But she was firmly told it was a question of respect for poor people who had no choice but worn-out clothes while she could afford to buy new ones. So she stopped wearing jeans with holes, but drew the line at ironing them. The English-speaking world can never reach those French levels of rational irrationality. Marder remembers walking home at night down the middle of the road to avoid the ubiquitous clumps of dog excrement on the pavements. To Parisians of the 1970s, it was of primordial importance to wear clean, chic, and well-ironed clothes, but the dirt all over their beautiful city could be ignored.

Finally, in her third year and third stage of adaptation, she felt well enough acquainted with French culture for there to be few surprises. Overall, she says, she left with an enduring love of the city, but without regret. Although everyone at the Ecole Normale was friendly and supportive, elsewhere she had felt awkward being a single American woman at a time when the French were both anti-American, particularly because of Vietnam, and not yet fully at ease with unmarried professional women.

The years in Paris certainly forced her to find her own internal resources. Perhaps when she went there she was no more committed to her career than most "kids" in their mid-twenties. Science is relentlessly demanding; for most of its practitioners, it becomes a vocation. I suspect she was struggling with that realization in Paris, and it must have been doubly hard because it came at precisely a time when her science was not going particularly well and she felt isolated. Somehow she went home with the commitment firmly made.

All in all, Marder's four postdoc years, 1975 to 1978, were rife with practical and personal difficulties, and her progress as a scientist was not immediately obvious. But this is emphatically not a wasted chapter in her life; it may not have led to new ideas in Marder's science, but it has influenced the way she does science—her later conduct as a leader in her field, her standards for herself, and her ambitions for her team.

I put it to her that she had in effect experimented on herself and in the end had been able to extract meaning from it. Marder agreed: "Right.

And that lesson was you must make your own mistakes; don't make other people's," by which she means that in a lab hierarchy, juniors can feel constrained to make scientific compromises, knowing that they shouldn't. She was sure all along that her concept was important and central to a new way of thinking about circuits; she now feels she should have pursued it with more determination and with more effort to convince her colleagues and superiors in the Paris lab. I told her I thought it had been hard for her. "It was. I made what I consider the best of a bad compromise, and I came away with lessons learned. It was nobody's fault. Probably I wasn't articulate enough at explaining why the cell identifications were so important to me. And you can't blame people for not understanding a paradigm shift before it's happened."

5 A Lab of One's Own

Arriving at Brandeis in autumn 1978, the new assistant professor found she had been given a perfect suite of small rooms, ideally suited to electrophysiology. It had not been in use for quite a while. The door opened onto some nine hundred square feet of litter, empty cardboard boxes, bulging garbage bags, and drawers full of tiny feet and other animal parts preserved in vials of formalin. Marder was appalled, but she rolled up her sleeves, threw the trash out, scrubbed and mopped, and started to acquire some equipment. Brandeis gave her a miserly setting up grant of $15,000. She couldn't even buy a full rig for that—just a microscope, an oscilloscope, and some manipulators—she had to use funds from her research grant to start equipping her laboratory.

Marder quickly made friends in a group of young faculty and felt at home on familiar turf. Shoehorned into the town of Waltham that encircles it, the Brandeis campus spreads in a seemingly haphazard manner over hilly grounds with shrubberies, gardens, and woods. It's pleasant because the buildings are modestly sized and carefully sited. It must have felt laid-back and casual after the formal cityscape of Paris.

Marder took an apartment in Cambridge and found that several old friends from graduate school were living there too. One of them, Ron Calabrese, a fellow neuroscientist in the invertebrate field, was also a newly minted assistant professor, at Harvard. They compared notes about equipment and organizing their labs, drank a lot of coffee, and retired to the movies at the end of their long days. Marder had met Calabrese when they interviewed for graduate school at Harvard in 1968. They had been "kind of pitted against one another," according to Calabrese. He was accepted but then turned Harvard down and went to Stanford. Over the following years, they had occasionally met because the world of invertebrate neuroscience was small and they went to the same seminars and meetings. Calabrese had always been a kindred spirit in recognizing the importance of circuit function; it was a relief for Marder to be fully understood and encouraged as she began her independent research.

Teaching, as she had foreseen, was not a burden. Her first undergraduate teaching assignment was a breeze; it was on synaptic transmission. The next was animal physiology. Marder had never taken an animal physiology course herself but was perfectly confident she could bone it up. She got three different medical physiology books and the course textbook, read the course material in all the books, and wrote her lectures. She thinks there was nothing in biology that she couldn't have taught at the time, adding ruefully that she would have had trouble if they'd told her to teach French. Marder still teaches an undergraduate course every year and every year finds the time to rethink each of her lectures. I have sneaked in a few times to see what's going on. She's a remarkable teacher; the flow of talking and drawing diagrams in a rainbow of colored chalks is relaxed but purposeful. Sometimes she'll relent and digress: "The problem with the description in your text book is these guys decided to tell you the truth. Now basal ganglion chapters never make sense, but this particular one doesn't make sense in a different new way. They used to tell you 'This is how it works' but here they admit they don't know how it works, so they talk about possibilities, which is fairly confusing and, of course, won't help you write assignments or answer exam questions." Or "Ah, yes, your assignments—some of you have very serious misunderstandings about toxins. Not good. Not safe. So listen carefully now."

With only four published papers on her record, Marder was eager to tie up the loose ends from her experimental work in Paris and add to her total. The paper describing neuronal responses to acetylcholine, gamma-aminobutyric acid (GABA), and glutamate written in 1977 in Paris with Danièle Paupardin-Tritsch had been published just after she arrived at Brandeis. Marder was disappointed with it, knowing it was limited in scope because the work had been done on unidentified neurons. It was a detailed, descriptive, pharmacological paper, with some worthwhile results but rather flat in tone; it didn't tell a story. It was, however, frequently cited, which is important in promoting a young scientist's name, because it contained a useful finding: a substance called pilocarpine, which acts on some of the acetylcholine receptors, reliably triggered rhythmic bursting in the ganglion, Thereafter, any researcher using that method to stimulate a ganglion listed her paper in the references.

They had planned a second paper that would look in detail at the responses of crab stomach muscles to acetylcholine, which of course sidestepped the problem of identifying neurons. Most of it had been written up in Paris, but some experiments were still missing.

In the Boston fish market, the local crabs were from the species *Cancer irroratus*, the Atlantic rock crab, and *Cancer borealis*, the Jonah crab, rather than the European *Cancer pagurus* she had found in Paris. At the time, the small but annoying differences in the neurophysiology of such closely related species were not suspected, and Marder's published paper merely notes that most of the experiments were carried out with *C. pagurus* in France and the remaining experiments with the Boston crabs. The paper, which Marder now qualifies as "profoundly boring," was written up and submitted to *Journal of Experimental Biology* just before Christmas 1979, four years after Marder had set out so nonchalantly for Paris.

With that unfinished business out of the way and with her new equipment up and running, she was free to throw herself wholeheartedly into the work that most deeply interested her. Although crabs were perfectly adequate for studying muscle responses and she could pick up a couple from the market on her way to work, to study the neurons of the stomatogastric ganglion, Marder wanted to go back to her old friend, the Pacific spiny lobster, *Panulirus interruptus*. She ordered some from California. They would have to be kept alive somewhere. She remembered Allen Selverston keeping his lobsters in a bathtub in a green-lit cold room; indeed, she well remembered hefting the carboys of water to refill it.

The Brandeis Biology Department was distracted by its own internal affairs that year, and no one seemed to have time to spare for Marder's predicament. The consensus was, "Oh yeah, well, put your animals in the cold room." It was someone else's cold room, of course, and it was already in use with the temperature set at 4°C—too cold for crustaceans. Marder installed some giant plastic troughs and pushed the temperature up a bit—too warm for the other residents, and they died off. It didn't take long for the salt water to corrode the chillers, much to the ire of the investigators on that floor. The stink and the leaks made a persuasive case when Marder went in for a Brandeis grant competition. She won enough money for a refrigerated seawater facility and real tanks.

In those days, before carrier services, the lobsters had to be put on a night freight flight, and Marder had to retrieve them from Logan Airport at whatever hour the flight arrived. There were a dozen lobsters in each box, and they would arrive squeaking and scrambling, the carton wobbling around and spooking the freight handlers who didn't want to touch it. Marder would drive the animals back to Brandeis, put them in the tank, and go home to worry about them. Usually they were fine, but it was an expensive business, and sometimes a whole shipment would die.

It occurred to me to ask why she bought spiny lobsters from California at evident expense, when *Homarus americanus* was shuffling around the Atlantic coast, on her doorstep. That, Marder told me, was a sort of scientific cautionary tale about the perils of following the herd. The stomatogastric ganglion preparation from *Homarus*, it seems, was always taken to be difficult to work with because it didn't have lively fictive motor rhythms in the dish. In fact, it was unresponsive because everyone had been putting the ganglion in a saline made with a recipe copied from Edward Kravitz's papers. He is an illustrious researcher, but it so happened that he had devised this saline purposely to keep *Homarus* muscle preparations from moving, which suited him in the 1960s when he was studying the neuromuscular junction. Somehow the recipe was handed on respectfully and believed to be the "right" saline for *Homarus*. "And that recipe has the wrong calcium to magnesium ratios and so everything is silent. The minute we flipped over to using the *Panulirus interruptus* saline, the *Homarus* preparation worked just fine. *Homarus* got a really bad rap just because of that wrong saline." Marder put her head in her hands in mock despair. "Until the late 1990s! Unbelievable."

* * *

For the first few months, Marder worked mostly alone. The early hires in a lab are important because the tone, work ethic, and team style quickly start to emerge. Marder, as a young principal investigator, would have to step into a new management role. From now on, she would face an unremitting demand for her time, mentoring, and management.

Her first academic hire was someone she already knew. Chris Lingle had been a graduate student with David Barker in Eugene while Marder was a postdoc there. His thesis work on neurotransmitter identification took Marder's findings in San Diego as a starting point and tied down some of the elusive points that she had not been able to clarify. He did this by going back to the neuromuscular junctions where Marder had found ambiguous results. Lingle showed that, as Marder had suspected, the neurons that innervate the extrinsic muscles, like the pyloric dilators (PDs), were cholinergic, whereas the ones innervating muscles intrinsic to the stomach all appeared to use glutamate. The particular neuromuscular junctions that had puzzled Marder turned out to be on muscle cells that have "extrajunctional" receptors—in the cell membrane but not under the neural endplate—and these were receptors for substances other than the transmitter released by motor neurons at the endplate. This showed that a muscle fiber could respond to more than one neurotransmitter; it could respond to indirect influences.

Lingle had then become interested in the biophysics of ion channel function and would ideally have liked to join a channel lab. But he had no good contacts in his chosen field. Marder had learned a lot about channel biophysics in Paris, but she was not at all a "channel person" and would not have been the obvious choice for him. Nevertheless, she could offer him a transitional step, and he came to her lab in July 1979. Lingle's knowledge of Marder's field was precious to her, and they did work and publish together, although much of the time he worked separately on channels. He was her first postdoc researcher, but Marder was already energetic and active on his behalf in a way that is now admired as typical of her. She introduced him to anyone she knew who might help him, starting with a Brandeis biophysics lab. Marder says, "He was exceptionally competent and independent, and knowledgeable, and got more so as he became more of a biophysicist. He used this period to completely reconfigure himself. He learned a lot from Chris Miller upstairs. He also was befriended by Paul Adams who had gone to Stonybrook and was very kind to Chris. Paul was always a bit of a renegade and liked to help smart people who didn't have pedigrees. So Chris had these sort of entrées into the channel world until he became well enough known on his own."

At almost the same time, Marder acquired her first graduate student. Judith Eisen had started her PhD work on the developmental biology of lobsters at the Woods Hole Oceanographic Institution. When her supervisor was killed in a road accident, his postdocs and students had to be placed in other labs. One of the white knights helping this orphaned group was none other than the distinguished Harvard neuroscientist, Edward Kravitz, "the hero of this story," Eisen calls him. "He arranged some interviews for me with various people, some much more famous, I mean Eve was just starting out, so she wasn't that famous at the time, although well known in some circles. When I met her, I was completely blown away by her intellect, the kinds of issues she was trying to approach. I was a baby graduate student coming into a new field, but it was just so clear to me that of all the people's work that I'd read, she had the really big ideas. I wanted to work with her."

Although Marder had not yet published much, Eisen had recognized the thoughtful assessments that are the hallmark of her writing and was inspired by her research plans. Eisen shelved her interest in developmental biology for the next three and a half years and joined Marder. For her part,

Marder considers Eisen's arrival her first piece of really good luck. "She was a dream researcher, extremely skilled, extremely smart, extremely talented, a total joy."

At this stage in the story, a piece of my own luck runs out: none of the lab books, after Marder's return to Brandeis, contains her questions, insights, guesses, or comments, only notes and summaries of experimental results. But these are bare accounts of what has been shown; there is no speculation on meaning, no reasoned proposals for the next steps, no clue to Marder's thoughts. Her writing has become legible and her notes are disciplined, her sentences complete, perhaps because she is now sharing the lab books with her team. By early 1980, although she was still fully active at the bench, her handwriting disappears, and the lab books are kept up by her collaborators. Still, I can see that in November 1979, she and Eisen were confidently identifying the individual nerve cells they were working on, and the lab books include cell identification drawings for most experiments. This was a relief and a source of satisfaction to Marder.

Marder says she set out with the explicit intention of passing on to Eisen everything she knew in the shortest possible time. Soon they were working in tandem, as equals. Eisen told me, "We spent a great deal of time together. You want to make a preparation last as long as you can, so we did experiments that went long into the night. And I'm guessing that probably none of her other students had exactly that experience at the bench. It can only happen at the beginning. It was true with my students—after a while you have enough people in your lab that they're teaching each other and then they figure out how to do things in ways that are different from how you used to do them. You're the principal investigator, and you take on other responsibilities, and you're only concerned with the data—unless something goes wrong."

With Lingle, they started more transmitter identifications on the basis of Marder's candidates and Lingle's recent work on the neuromuscular junctions. Marder walked through the lab alone late one night and saw that there were three rigs in use, three experiments running, and thought, "It's a real lab!"

* * *

The stream of results that flowed from her collaboration with Eisen was so important in Marder's career that I want to describe the experiments in more detail than I intend to provide for later ones. These experiments allowed Marder to articulate the concept of the neuromodulation of circuits. The phenomena that Marder and Eisen observed, and the insights that stemmed from them, profoundly changed neuroscience by revealing the flexibility of neuronal circuits. It was significant work and applicable to the circuits of all nervous systems. Of course, in 1980, they were not alone in trying to explain the activities of neurons. Several other labs were looking at the electrical properties of single neurons and the modulation of synapses or single currents, establishing the concept of the variable excitability of neurons. It was becoming apparent that neurons could no longer be thought of as on-off components in a chain of information, immutable in their properties, but had to be appreciated as dynamic, with a constantly changing responsiveness influenced by their environment. Marder, however, had always been convinced that neurons should be studied in the context of the circuit. Now she was putting together the principle of exogenous neuromodulatory control of circuit dynamics. She also developed her insight into the functional purpose of those modulatory phenomena: they could reconfigure a circuit in a way equivalent to rewiring it to produce variations of behavior.

Everyone takes neuromodulation for granted now. It is completely accepted as a key aspect of how all nervous systems work, and the textbooks have been rewritten. But in the early 1980s, putting the evidence together and recognizing its implications produced, in the time-honored and worn-out cliché, a "paradigm shift" in thinking about neurons and networks.

Although earlier in her research career, Marder already had the intuition that neurons might be susceptible to influences from their environment as well as responding to other neurons with which they had direct synaptic contact, so far she had found no obvious experimental approach for her investigation. Marder and Eisen therefore began by addressing the research question in Marder's grant application. Were the electrically coupled anterior burster (AB) and pyloric dilator (PD) cells using the same neurotransmitter? Electrical coupling depends on "gap junctions," a good descriptive term for the electrical synapses where two cells have membrane channels closely facing each other that allow small molecules, including charged particles, ions, to pass from one cell to the other. The ions are attracted down a voltage gradient, in effect transmitting current across the narrow gap. This is obviously a faster method of communication between cells than the secretion of neurotransmitter molecules by one neuron into a synaptic cleft to activate receptors on the receiving cell. It was thought that speed was useful for rapid escape mechanisms or for keeping cells synchronized, but otherwise their existence alongside chemical neurotransmission was an enigma.

In the late 1970s, at the time of Marder and Eisen's experiments, it was emerging that gap junctions were far more than an occasional curiosity; they were widespread in all nervous systems, but research on their structure and myriad functions had not yet answered some fundamental questions. It was, however, known that unlike chemical synapses between neurons, gap junctions could be bidirectional. Given that the PD cells used acetylcholine, if the AB cell used glutamate, could there be trafficking back and forth of acetylcholine and glutamate? What would keep a glutamate-using neuron's characteristics stable if it was getting acetylcholine at all its gap junctions? Would that threat dictate that the two cells would be constrained to use the same neurotransmitter? On balance, Marder expected they would be. Finding out, however, was not a simple matter, and the quest got fruitfully sidetracked early on.

To understand the state of affairs from which they were starting, we must look at the connection diagram for the pyloric circuit as it was known to Marder in 1979 (see figure 5.1).

Although most of the connections that Don Maynard had mapped out in the early 1970s could be relied on, puzzling results in some experiments left researchers struggling to fit their observations to the diagram. The AB and two PD cells were shown as a group, and this was the source of troublesome ambiguity. It was known that the AB cell was electrically coupled

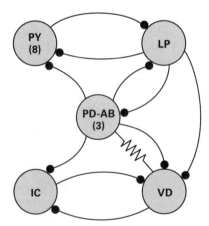

Figure 5.1

The synaptic connections of the pyloric circuit as accepted in 1979. The PD-AB group was treated as a single unit. Black circles indicate inhibitory chemical synapses. The resistor symbol indicates electrical coupling. Neurons: AB, anterior burster; IC, inferior cardiac; LP, lateral pyloric; PD, pyloric dilator; PY, pyloric; VD, ventricular dilator.

to both of the two PD motor neurons and that the three neurons fired in synchrony.

Marder knew from her work in San Diego that the PDs used acetylcholine in their excitatory connections to the dilator muscles of the pylorus. But the AB-PD group was known to inhibit follower cells within the ganglion. Was it possible that acetylcholine was eliciting inhibitory postsynaptic effects inside the ganglion at the same time it was exciting the muscles outside it? Testing for acetylcholine was not feasible because the molecule is so unstable, but substitutes, such as carbachol, excite the same receptors.

More difficult was the question of the mysterious AB cell's neurotransmitter. The AB is an interneuron; it makes all its chemical synaptic connections in the tangled neuropil of the ganglion. It is the only interneuron among the fourteen neurons in the pyloric circuit of the stomatogastric ganglion. The others are motor neurons that extend axons in nerve fibers to muscles, where any effect is easy to measure. The paired PD motor neurons are large neurons; the AB is small. Because the AB cell is small and has no muscle target, it was difficult to identify its neurotransmitter. Glutamate was a strong candidate, again from Marder's earlier work; however, glutamate is an amino acid found in all animal cells as a metabolite so its presence in a neuron certainly wouldn't prove it was being used as a transmitter. For this problem, Marder now held a trump card: her work in France had shown that picrotoxin, which was known for its ability to block $GABA_A$ responses, could also be used to block the stomatogastric ganglion neurons' inhibitory responses to glutamate. The lateral pyloric neuron (LP) is inhibited during the AB and PD cells' bursts, and it has a synaptic connection from the AB. If picrotoxin blocked the LP's response to AB and PDs, then it would show that glutamate was involved. If atropine blocked it, then it would show that acetylcholine was involved.

From the start, Marder and Eisen were hampered by frustrating results. They did the picrotoxin experiment again and again, and sometimes the LP response was blocked and sometimes not, and—most infuriating—sometimes it was blocked a bit. They checked every element of the experiment and tried again and again. Marder remembers imagining they were facing into a corner, with nowhere to go.

Staring at the wiring diagram, Marder and Eisen realized it wasn't the whole story. It showed that the grouping of the AB and the two PDs released transmitter onto the LP, which was inhibited by them. But trying to identify those transmitters by using pharmacological blocks was doomed to failure because the transmitter they were blocking could be coming from both the AB and PD cells or only one or the other.

It suddenly dawned on them that if the AB and PD cells used different transmitters, then the inhibition of the LP could be caused by two neurotransmitters. What happened if the LP received more of one neurotransmitter than of the other? That would occur if either the AB or PDs were more excited than the other—releasing more neurotransmitter and evoking stronger inhibition in the follower LP cell. Thus, if the ratio of the two transmitters being released was not fixed, then, as their experiments were blocking responses to only one transmitter, the result would depend on the relative amounts of the two transmitters that had been released and would give the inconsistent results that had flummoxed them. In understanding the subtlety of neurons, the idea was revelatory.

Marder learned a lesson she has continued to teach to this day: "Whenever something changes when you're doing the experiment the same way, and you check you're doing it right, and you still get really difficult to understand answers, it's because there's something important hiding in there. Experiments that apparently fail give you new insight."

The explanatory idea was in itself a breakthrough, but it was an unproved supposition; above all it showed that the methods of classical neuropharmacology were not adequate in this case. Marder decided to try one of the new techniques for knocking out a single neuron while keeping the rest of the circuit intact. For the first time, investigators could see the role of each neuron in the circuit, its connections, its function, and its properties, by elimination. One of these techniques was photoinactivation, and Selverston, with a graduate student, John Miller, had just perfected a method using a dye with the dashing name "Lucifer yellow."

Those were pioneering days for such experiments: a petri dish was prepared by drilling a small hole in the center and sticking a piece of a glass microscope slide over it. The dissected ganglion was placed over this glass window. The dye was injected through a microelectrode into the neuron to be eliminated, and an intense blue light was focused onto the ganglion from underneath the dish. The light activated the dye and killed that one neuron. Today the method is much easier; microscope attachments can aim focused fluorescent beams directly onto a preparation.

In Marder and Eisen's experiments, the traces of activity recorded when the AB and both PD cells were alive showed the inhibitory influence of the LP cell's firing. The PD cells continued to show this influence after the AB cell had been eliminated. But when the PD cells were killed, the LP cell had no inhibitory effect on the AB cell (see figure 5.2).

Now the actions of the AB and PD neurons could be separated, and Marder and Eisen found that the inhibition of the LP caused by the AB cell

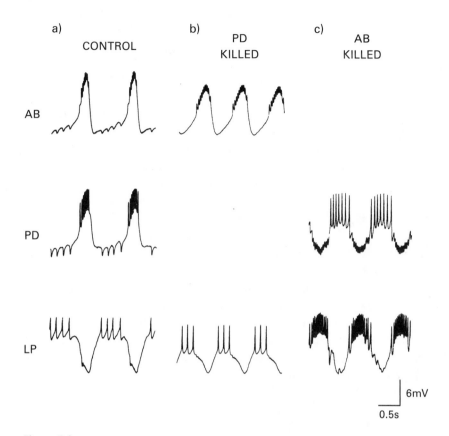

Figure 5.2

Examples of traces of simultaneous intracellular recordings indicating the effect of photoinactivation of neurons in the AB-PD-LP group. (a) At left, shows the control result: the AB and PD neurons burst in synchrony, inhibiting the LP neuron. The LP neuron's action potentials affect the AB and PD neurons, causing inhibitory potentials that appear on the trace as downward spikes between bursts. (b) Center, when the PD cells have been photoinactivated, bursting of the AB cell continues to inhibit the LP cell, but the AB cell no longer shows any effect from the LP cell's action potentials. (c) Right, when the AB cell is inactivated, PD cell bursting continues to inhibit the LP cell. But here, each action potential in the LP's bursts caused inhibitory potentials in the PD neurons. AB, anterior burster; LP, lateral pyloric; PD, pyloric dilator. (After Eisen and Marder, 1982.)

was indeed blocked by picrotoxin, not atropine, indicating glutamate. This showed conclusively that these two neuron types, AB and PD, although electrically coupled, use different neurotransmitters. It also answered one of Marder's questions in 1974 at the end of her thesis: PD neurons did in fact use a single neurotransmitter, acetylcholine, at both their excitatory connections with muscle and their inhibitory connections in the neuropil. Henry Dale was vindicated!

Photoinactivation also resolved the longstanding puzzle of the ganglion's continued rhythmic activity in the dish. It would continue to produce its fictive motor patterns after dissection as though it were still controlling muscle and, at least in the spiny lobster, usually do so even if the "front end" (the esophageal and commissural ganglia) were cut off. In that case, there could be no further stimulating signals reaching the ganglion from neurons outside it. So one or more of the neurons in the ganglion had to be the "pacemaker" that initiated and orchestrated these cycles of neuronal activation in the isolated network. In San Diego, Selverston and Miller were able to use photoinactivation cell kills to determine that the AB neuron alone was the master regulator. By the summer of 1981, Marder and Eisen had replicated this work: in June, Marder wrote to Brian Mulloney. After her signature comes a laconic postscript: "Oh, forgot—the AB is probably the burster—the PD doesn't burst when AB is killed—Miller & Selverston—we confirm."

* * *

One windy day in the middle of that summer, Marder got herself some lunch at the students' union and sat in the sun. She was in a terrible mood for no particular reason, and the wind kept snatching at her newspaper, and then part of it blew away. A well-dressed man picked it up and brought it back to her, which somehow irritated her. So she was ungracious, but he went on to ask whether she was on the faculty, and she just, but only just, let him know that she was.

"I am too," he said.

And Marder almost groaned, "I figured that."

He asked, "What do you do?" When she said neuroscience, he gamely said he was interested in the brain too. Marder was by now longing to freeze him out, but he was faculty and by a few years her senior, so she felt cornered and agreed to a beer or a coffee sometime. He said he would call her.

Professor Wingfield had spent the previous year on sabbatical at Cambridge University in England. He was investigating speech, language, hearing, and memory. In short, he was a cognitive neuroscientist, except that the profession hadn't yet been constituted. When they had that beer together, he asked if he might sit in on her neurobiology course in the autumn.

And that is how Arthur Wingfield came into Marder's life. He took the whole semester, sitting at the back of the class with a wildly good-looking party guy he and Marder nicknamed Steve the Surfer (Marder can't remember why). Fully appreciating the cut of Wingfield's English tailored jackets, Steve stood out himself for his expensive clothes. He had been sent to Brandeis to get a decent education and take demanding classes, but he was no intellectual, and his father had bought him an apartment and given him too much money. However, all being well that ends well, in due course, the young man became a successful and wealthy businessman, and Marder says she still enjoys hearing from him now and then. This unlikely pair became good friends, and Wingfield thus found out that all the students were worried he would take the exam and break the curve. Wingfield let it be known that he didn't find public humiliation appealing and no way would he take the exam. Most of the semester went by with Wingfield and Marder exchanging a word now and then after class. Eventually they started having the occasional beer or coffee. By that time, Marder felt strongly that he had an unfair advantage over her. "The students in your class know you better than you know them. It's a very asymmetric relationship because you're performing in front of them, and they're not."

I have it on the good authority of several of her friends that Wingfield was, and is, entirely suitable for her. At the beginning of 1982, he moved into the 450-square foot apartment on Harvard Square that she loved. He parked his bigger possessions in his lab, but it was still a tight fit. A few years later, another tiny apartment next door came up for sale and they bought it, using the connecting balcony between the kitchens rather than knocking a hole in the dividing wall because it was an old building with idiosyncratic plumbing and uncertain structure, and anyway they moved before they got around to it. And they didn't get around to marriage until 2001.

* * *

1981 and 1982 passed in a flurry of experiments using photoinactivation, electrophysiology, and pharmacology. Marder and Eisen got a spate of results that I can only summarize in this chapter. No conceptual breakthrough fell easily into place, and they weren't sure what they were looking for. They moseyed their way through a lot of experiments, accumulating many disparate bits of information. As Marder describes it, "We were just following our noses at that point."

The revised wiring diagram separated the AB from the PD cells and showed feedback from the LP neuron inhibiting both the AB and the PDs, feedback that other researchers had reported detecting. Marder and Eisen found that the inhibition in the AB neuron seemed to be slower and weaker than that in the PDs, but they couldn't pin it down. Once again exasperated by ambiguous results, they started drawing all the possible synaptic connections of the AB-PD-LP triangle that might explain what was going on. There were nine of them (see figure 5.3).

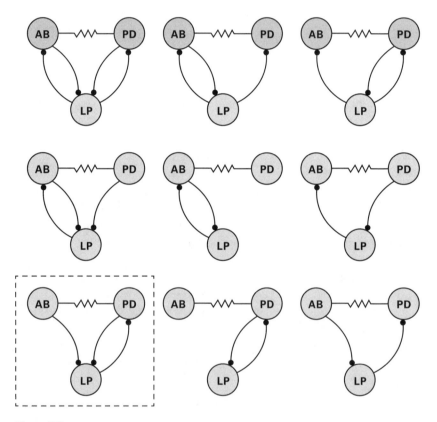

Figure 5.3
The nine possible diagrams of synaptic connections between the AB, PD, and LP cells. Top left shows the generally accepted diagram in 1981. Marder and Eisen showed that the connections at bottom left are correct. Black circles indicate inhibitory chemical synapses. The resistor symbol indicates electrical coupling. Neurons: AB, anterior burster; LP, lateral pyloric; PD, pyloric dilator, of which there are two. (After Eisen and Marder, 1982.)

Marder and Eisen set to work on a process of elimination. They discovered that when the PD cells were killed, action potentials in the LP had no effect at all on the AB neuron. They concluded that the LP cell made a synaptic connection to the PD cells but none to the AB. The AB appeared to be inhibited only by secondhand feedback: inhibition from the LP was passed on by the PD cells through the electrotonic coupling. The wiring diagram all stomatogastric ganglion labs were using, which showed an inhibitory connection from the LP to the AB, had to be corrected. Now it would show that all chemical influences onto the AB must come from substances secreted as hormones or released by neurons outside the ganglion. Thus, the vital pacemaker cell can respond to the state of the animal but is buffered against strong feedback from the circuit, protected from excessive inhibition and conflicting signals.

Then Marder and Eisen discovered that the AB and PD cells differed, not only in the neurotransmitter each used, but also in their responses to the presence of substances such as dopamine that were known to reach them in the neuropil. The same message relayed to the ganglion by incoming nerve fibers could be interpreted differently by the two types of neurons. Marder and Eisen saw that the presence or absence of a modulating substance could differentially affect the activity and thus the amount of transmitter released by one or the other cell. This result neatly placed another piece of the jigsaw puzzle, showing a functional purpose for their finding that the effect of the AB and PD cells' firing on the follower LP cell depended on how much transmitter each had released. The two types of cells responded differently to the same environment.

After that, they observed that the inhibitory effects caused by the AB and PD cells were not the same, although they fire synchronously. The AB cell using glutamate caused a fast postsynaptic potential (PSP), while the PDs using acetylcholine caused slower PSPs. The PSP evoked in the follower neuron was the summation of the simultaneous action of two different neurotransmitters on different receptors. They realized that if a follower cell received more acetylcholine relative to glutamate, the response of the follower cell would be retarded. This is called a phase shift. So a modulatory substance that increased the PD cells' release of acetylcholine, but not the AB cell's release of glutamate, could cause a phase shift in a follower cell. A group of cells that had been thought to act in concert turned out to be composed of soloists with different talents and sensitivities and thus the potential for different responses to their shared environment.

Another unsuspected detail Marder and Eisen revealed was that the frequency of the pyloric rhythm was modulated independently of any phase shifts. The frequency seemed to be controlled by the AB neuron alone, while both the AB and the PD cells could alter the phase relations.

Putting all their evidence together, Marder and Eisen were able to state that in response to the presence in the neuropil of different substances, a subcircuit of neurons could alter its rhythms. That led them to think about the functional consequences for the circuit. "But we were led there; we didn't go looking," Marder insists. After the first results that prompted them to look more closely at the actions and characteristics of the AB and PD cells, the series of experiments may well have been unplanned, with each result suggesting the next test, but the links in the chain of experiments needed were becoming clear, and there was a heady rush of work. They were constantly surprised and excited and often in the lab until late at night. Eisen says she tends to recall only the exhilaration: "One of the really nice things about the experiments we did—electrophysiology—was that they gave us instant gratification. You could see the result coming off the chart recorder, and we had an audio monitor set up so we could hear it even if we were in the other room. We'd call out, 'That's exactly the effect I expected!' or 'Wow, look at this, this is weird'."

As they worked, they discussed the presentation of their results; they wanted each discovery to be related to the context of the pyloric circuit. Eisen remembers, "We would mock up the papers and get a sense of what the story would be, and what other experiments we needed. Sometimes we'd be doing an experiment and writing at the same time—there's a lot of downtime. We had all these different traces, on long rolls and then we had to cut them out and paste them into notebooks, and if you didn't want to cut it some place important you folded it up, and a long experiment might end up filling more than one notebook."

It sounded a lot like kindergarten to me: all scissors and glue. Eisen laughed, "That's exactly right. To me there's a kinesthetic part of science that's totally lost with the beautiful pictures on people's computers, and it makes it a lot harder to look at your students' stuff than having those communal lab notebooks that you pulled down from a shelf to look through the data."

In late 1981, they submitted a report of some of their results that was published in 1982 as part of a suite of four papers on work using photoinactivation. The report is detailed and assured in tone, but the discussion section is hedged with circumspect phrases. It was not until the summer of 1983 that Marder knew they had "the big story." They wrote up the work in three articles that came out together in 1984 in the same issue of *The Journal of Neurophysiology*. Such back-to-back publishing of more than one related piece of research by the same authors was not rare in those days, and it allowed for a measured setting out of the evidence and for building up a strong statement of its implications. Marder and Eisen took the reader through their results, pointing out the functional utility of each and concluding, "We are beginning to define the mechanisms by which a small number of neurons with anatomically defined sets of synaptic interactions can produce a wide range of motor outputs."

Marder had decisively answered her own question from her lab book in 1971: "Is it possible to make any guesses about whether it is a useful model to think of a small nervous system as a circuit of equivalent units, or in fact do the units have to be individually characterized?" Indeed they do; each "unit," each type of neuron, has its own individual, intricate, but malleable character that must be analyzed and observed in context. In a circuit, no neuron is just an on-off link in a chain. Nothing could be taken for granted. She had been right all along in her insistence that only the study of identified neurons would be valid because, by studying exactly the same cell types again and again, she had been able to recognize that each has multiple personalities, not a fixed phenotype. Each cell type has a repertoire of possibilities that are expressed under different conditions of neuromodulation, that is, in the presence of the various modulators, released as hormones or by other neurons as a result of the animal's changing circumstances. In this way, neuromodulation can adjust the interactions among neurons in a circuit without any change to the anatomical connections of the system and without change to the number or strength of synapses. The big, new story was that, through neuromodulation alone, circuits of neurons could reconfigure themselves to vary the circuit's output and hence the animal's behavior.

Eisen left the lab to take up her postdoc position—in Eugene, Oregon, where she is now a professor. Few young scientists can ever have graduated with such a game-changing doctoral thesis. Her bold choice of Marder as her PhD supervisor, on scant evidence, had been rewarded.

For Marder, the "tenure clock" had been ticking away, and she now had to pay attention to confirming her position at Brandeis. With a convincing body of work and three major papers about to be published, Marder became an associate professor in 1984. She wrote a concise article, putting the findings of the last three years together under the title "Mechanisms Underlying Neurotransmitter Modulation of a Neuronal Circuit" that was published in *Trends in Neuroscience* in early 1984. Her long-held intuition about neuromodulation was now clearly formulated, with examples backed by solid data. Her position in the vanguard of invertebrate neuroscience was beyond doubt. Here, it is regrettable but necessary to qualify "neuroscience" with the word "invertebrate." In the world of vertebrate neuroscience, resistance to accepting the relevance of results and principles coming from invertebrate models was always widespread. As Allen Selverston had written irritably in 1976, "There is often criticism of such simple networks on the grounds that invertebrate neurons are different, both at a structural and network level, from those found in the more complex nervous systems of mammals. ... With respect to basic differences in the functioning of invertebrate neurons, it is really the responsibility of vertebrate neurophysiologists to show how *their* cells are different from those found in simpler systems." They couldn't, of course, and never did, but acceptance was gradual, took decades, and even today the myth lingers on.

* * *

Marder had first asked herself "What are the transmitters in the stomatogastric ganglion?" in 1971. She was referring to its thirty neurons and assuming that they communicated with each other as shown in the 1971 wiring diagrams, each using one specific neurotransmitter. It transpired that those thirty neurons use only two neurotransmitters between them, acetylcholine and glutamate. As the abundance of molecules that didn't act at the distinct neuron-to-neuron synapse but were nevertheless active in nervous systems was gradually revealed, she asked herself why there were so many of them when nature generally has to use resources sparingly.

In Oregon and Paris, Marder's lab notebooks show her recognition that the responses and interactions of neurons were likely to be complex. David Barker's team had identified substances present in input nerve fibers to the stomatogastric ganglion but apparently not found in its neurons. He thought of one of them, octopamine, as a circulating hormone affecting the strength of neuromuscular junctions, those special synapses between motor neurons and muscle. But Marder had realized that a circuit of neurons, such as the pyloric rhythm-generating circuit, being more intricate, would reveal more about modulation than the relatively simple neuromuscular junction.

She had speculated to herself that, as well as being affected directly by neurotransmitters, neurons might also be influenced—primed or damped—by other substances. The substances might surround the neuron, perhaps hormones or other circulating molecules. Or perhaps they were released into the neuropil by neurons that were not even connected to the one in question. Thus, neurons in a network would be getting at least two kinds of message: direct from synaptic contacts and diffuse from substances in their fluid environment. This second sort of influence would be active on receptors distributed all over the neuron's membrane rather than at synaptic clefts. Receptors are specialized for particular molecules so this meant that a neuron might be able to respond to many different substances; it just depended what receptors it had. Kehoe's research had opened up a new set of possibilities by showing that there were three different receptors for the single neurotransmitter, acetylcholine, and that each responded with its own distinctive time course. It was clear that the action of any one substance was not necessarily the same everywhere but depended on the array of receptor types on each neuron. Perspicaciously, Marder had been thinking that, in addition to the presence or absence of receptors for specific substances in a neuron's membrane, the composition of that "population" of receptors would be interesting to investigate. How many different types of receptors might a neuron have, and how would the neuron control the total number and placement of each type?

Marder was also thinking about the consequences for the small system she was studying of receiving not one but several modulatory substances; they would enable a wide range of precise, specific adjustments to the system's functions. She saw that the varied responses of different types of neuron, as they were differentially modulated, would enable their network to respond flexibly. It could change its output, running muscle movements faster or slower, for example, according to incoming information about the lobster's feeding and the state of its digestive processes.

The concept of neuromodulation of a circuit brought a completely new dimension to the study of neurons. Marder now knew that if she ever wanted to understand the workings of the stomatogastric ganglion, she would have to identify all of the modulatory substances influencing it and describe their effects. She had no inkling of the profusion of such substances that would be found in the next decades.

* * *

In that vintage 1981 class, with Arthur Wingfield and Steve the Surfer, a graduate student named Scott Hooper came to Marder's attention. He was, Marder says, "wicked smart," and certainly his track record was impressive. He came from MIT with two undergraduate degrees, in physics and biology. After that, he didn't know what he wanted to do, but doing nothing was contrary to his family work ethic, so he got a job as a laboratory technician at Brandeis. It was immediately obvious to his bosses that he should be in a PhD program. Among his courses, he took Marder's neurobiology, and he also did the customary rotations, one of them in her lab. He stayed on. Eisen was finishing off her doctoral thesis by then. Hooper's charm persuaded the two women to help him with dissections, and he was disappointed when this indulgence ran out. Marder adds, "Early in his career, Scott was afraid of wrecking expensive lobsters—that made him nervous, so he liked the crab and felt freer with it." Soon enough, Hooper's own talent and drive took over. "When you look through the microscope and see the forceps, you go into a meditative state—a connection of mind to fingers—can you get it all perfect?"

Marder thinks she was fortunate to have Hooper in her lab when her brilliant, friendly collaboration with Eisen had to come to an end. "Scott was one of the most exciting, original, and creative people I've ever worked with." It was a sensitive transition, and it coincided with her growing interest in the identification of neuromodulatory substances and the analysis of their specific effects.

By the early 1980s, the substances reaching the neuropil, either in the hemolymph surrounding the ganglion (which lies in an artery) or secreted by axons running in the stomatogastric nerve into the neuropil, were thought to include the amines: dopamine, histamine, serotonin, and octopamine. At Harvard, Kravitz's team had been studying these amines in various parts of the lobster's nervous system. The evidence that they were also present in the stomatogastric ganglion, much of it acquired in David Barker's lab, was found by fluorescent methods in the nerve fibers entering the ganglion. Further evidence in the form of electron microscopy images from the lab of Teddy Maynard showed that these fibers contained neurons with the type of large "vesicles" that secrete such substances. It was also suspected that various peptides were involved, and Marder wanted to investigate those too. (Although a much-used noun, "peptide" has never been precisely defined. It refers to a string of amino acids, more than one, but not enough to be called a protein. The borderline hovers around fifty.)

At the 1982 Society for Neuroscience meeting, Marder ran into Ron Harris-Warrick, who had been one of Kravitz's postdocs at Harvard. He was starting his own lab at Cornell—in fact, it was the job that had been offered to Marder, and he started there the year after she went to Brandeis. He had been working on serotonin, and by the end of their conversation, Marder had persuaded him that the stomatogastric ganglion was the ideal preparation for him. He visited her lab with his first graduate student, and they spent a week with Marder and Eisen learning the dissection and recording techniques. Then Marder noticed a paper on identifying serotonin in the lobster nerve cord. It relied on immunohistochemistry and came from Barbara Beltz, another postdoc in Kravitz's lab. "And so I said to Barb, 'Wouldn't it be fun to try the stomatogastric ganglion?' I dissected some out, we fixed them, and she ran them. Eventually she taught me how to do the immuno, and I did it here myself. That first serotonin stained beautifully, it was amazing, it was just gorgeous."

The upshot was a paper written with Eisen, Hooper, Beltz, Harris-Warrick, and Bob Flamm, his student. They made a comparative study of three species—the spiny lobster, the clawed lobster, and the rock crab—using immunohistochemistry to show which regions of the stomatogastric ganglion system contained serotonin. The electrophysiological results were even more interesting. It is, I think, the first demonstration of different forms of neuromodulation in the ganglion. In all three species, the pyloric rhythm was affected by serotonin, just dumped, as Marder used to say, onto the ganglion (it's officially called "bath application") but at quite different concentrations. In the spiny lobster, there was a response at the low concentration range typical of naturally occurring blood-borne hormones. However, the other two species required three orders of magnitude higher concentrations before responding, which suggested that in the living animal, the serotonin was delivered to the ganglion by the incoming nerve fibers.

A competitive situation could have arisen between their two labs, but Marder and Harris-Warrick agreed on a division of effort that lasted for several years: his lab would continue to investigate the role of the amines in the stomatogastric ganglion, while Marder's lab would take on the first studies of peptide actions there. Marder explains that Harris-Warrick had a longstanding interest in amines and the related signal transduction mechanisms, and so it was logical for him to continue working on the amines. Then she adds, a touch slyly, "Besides, it struck me that the peptides were all new territory and more interesting."

Marder set Scott Hooper to look for peptides that might influence the ganglion. He started, in the lab's customary style, just following his nose and quickly got interesting results with a small peptide called proctolin. "One day, I looked in the freezer. Eve had bought all this stuff, and I just grabbed something—proctolin! That work became my thesis. Years later she said, 'You're telling me you chose at random? I thought you *meant* it'."

Selverston's lab had already tried proctolin in the stomatogastric ganglion but hadn't seen any effects. Hooper succeeded because he happened to be cutting the front end off—the commissural and esophageal ganglia. That dampened the state of the ganglion, and it turned out that proctolin's effects are too subtle to be picked up if the ganglion is running strongly. Its effects are state-dependent, and only if the ganglion is quiet will proctolin noticeably rev it up.

Marder and Hooper wrote a short paper reporting that both proctolin and another peptide, FMRFamide (a mouthful, but the name just refers to the sequence of amino acids forming the peptide), acted on the stomatogastric ganglion, but not in the same way. It was an important step in developing the notion that different modulators would affect the network in different ways. Marder was wondering just how many modulators there might be. "Once we had two peptides and we knew the amines were there, I wanted to know what else. So there was a lot of what-elsing for a while."

But first it had to be shown that these peptides were relevant to the ganglion—that it was not an accident that these two substances, perhaps foreign to the ganglion, just happened to have an effect on it. After the success of the serotonin immunohistochemistry, Marder felt encouraged to use the same method to look for evidence of the two peptides in the neuropil of the ganglion. The first efforts showed that "proctolin-like" and "FMRFamide-like" immunoreactivity was found in input nerve fibers to the neuropil in the rock crab. It was clear that fibers in the neuropil showed staining but the stomatogastric ganglion neurons themselves did not. However, the exact identity of the peptides the antibodies the lab was using had bound to could not be confirmed using immunohistochemistry alone.

For immunohistochemistry, an antibody or antiserum is "raised" by an immune reaction in a laboratory animal, such as a rabbit, exposed to the target substance, in this case a peptide like proctolin. When the antibody is put on an experimental tissue sample, it binds to the target substance if it is present, showing where it is localized. (A secondary antibody, often tagged with fluorescence, is used to bind to the first so that it can be picked out under the microscope.) Two problems present themselves right at the starting block: one is to purify the peptide itself, from animal tissue, for use as the target substance against which to raise the antibody. The other is to make an antibody that will bind to that peptide specifically or relatively so. Many, many antibodies turn out to be unspecific and bind promiscuously to all sorts of similar molecules.

Marder turned to the catalogs and ordered any promising antibody she saw. Not many were available, however—very few for invertebrates and almost none specific to crustaceans. Some antibodies were raised by other scientists but not commercially available. The antibody she used for proctolin was not for the peptide isolated from crustacean tissue, although it was at least from an arthropod. FMRFamide was more of a poser; the only obtainable peptide had been isolated from molluscs, and although it had an activating effect on the ganglion, Marder knew it was probably not the endogenous peptide found in her species.

Marder learned to stick with peptides that came from other crustaceans like the red pigment concentrating hormone (RPCH) that was first isolated from shrimp; it worked on the stomatogastric ganglion like a charm, first time, whereas any number of molluscan peptides didn't really work. "It was very clear to me at the time that the problem with just getting peptides from other people and other places was that while the antibodies might work, often the receptors really did care what the precise sequence was." The antibody might find a similar molecule and bind to it, but then the peptide, in follow-up electrophysiological experiments, might fail to produce any effect on the ganglion.

All the same, it was obvious that peptides had a hugely important role in modulating the neurons of the stomatogastric ganglion, and they are now known to be the biggest group of neuromodulators in that system. It was also obvious that the obstacle to accurate work at the time lay in identifying and then purifying exactly the right peptide for the right species. Marder put a great deal of effort into getting it right. "At one point, I spent time over at Ron Calabrese's lab at Harvard. We purified the extended family of FMRFamides from the crab. But that meant getting brains and thoracic ganglia from hundreds of animals, grinding them up, and then doing high-pressure liquid chromatography assays in every fraction to separate the component substances, then doing inhibition-of-binding curves. And all of that *before* we made the antibody. Of course, we had to characterize the specificity of the antibody, which meant doing tons of radio-immuno assays, and so it was a year of work to make one antibody and characterize it—*after* you thought you had a pure peptide. It was a long haul in those early days."

The long haul went on for years, and more "novel" peptides were slowly and painstakingly identified in the stomatogastric ganglion. Relief finally came in the form of mass spectrometry, a technique that analyzes the component molecules of chemical compounds by their mass and electrical charge. At a meeting in 1999, Marder heard a talk by Jonathan Sweedler about his work on sea slug neurons using MALDI,* a new type of mass spectrometry that freed researchers from the labor of purifying peptides from tissue. The structure of a peptide could be worked out and the exact peptide synthesized. Naturally, Marder jumped at the idea; she is always interested in new techniques and unerringly picks out the ones that promise to

---

*Matrix-assisted laser desorption/ionization.

answer some of her questions. She set up a collaboration with Sweedler, and he assigned a graduate student, Lingjun Li, to work in collaboration with Marder's lab. MALDI was a transforming technology, allowing an immediate and astonishing glimpse of the sheer numbers of substances reaching the ganglion. From its tissue, Marder says, "We got a giant spectrum with 150 peaks. We knew what only 3 or 4 of them were—proctolin, for example. So the first MALDI runs were just *tantalizing* because all we could do was try and identify on the spectrum the few substances we already knew were there."

Li saw that Marder had a store of intriguing physiological questions aimed at uncovering the rules of neuromodulation. She went back to the Marder lab as a postdoc and has since carried on in her own lab, spending a good fifteen years working out what all those peaks are. Li has discovered more than two hundred novel neural peptides, more than twenty of them are found in the stomatogastric ganglion. (See plate 6.)

Some of them occur in "families" of related forms, such as the FMRFamides, and twenty or thirty members of a given peptide family may be found in any one crustacean species. This is an embarrassment of riches, Marder laments. "We went from having immunoreactivity with no knowledge of the physiology to having these giant families of peptides all of which are probably biologically active. It becomes quite a challenge. If Lingjun gives me the structures of 20 closely related versions of a peptide and it costs a thousand dollars to synthesize each one and then somebody's got to test them in a bioassay, it's very demanding. And you have to decide what you're going to do with all that information. So generally we go after a couple of them, perhaps the ones that look the most different, but that doesn't mean the others aren't there and doing things."

Of course, this raises the question of what such a multitude of peptides might be doing in the ganglion; what is their functional significance? Why are there so many different versions of the same peptide in the same species? Why, even more exasperating, do different species have different members of a common peptide family? Every crustacean species has many FMRFamides, for example, but some of the FMRFamides are found commonly, whereas some are species-specific. In contrast, some peptide families have small numbers or no variations, like proctolin, which is found in all crustaceans as the same molecule. Marder says, "We don't really know what to make of it. It's deeply annoying because it's so difficult and expensive to synthesize and test each one. There are some very interesting biological questions in there, but it's not easy."

Are some of these similar molecules redundant, or evolutionary vestiges, or doing a precise job? It is a formidably perplexing picture. Marder doesn't think it's degeneracy per se but rather that duplications build up on the peptide's gene, multiple copies in a row. She acknowledges not knowing how influential these different versions of a same peptide may be—or how trivial.

Research in Marder's lab has shown that two versions of a peptide can have similar effects on a given neuron, as though the neuron's receptors didn't distinguish between them. But experiments blocking the peptidases—enzymes that break down those particular peptides—showed that one of the two versions was apparently broken down much faster than the other. This could show that they have different affinities, one of the versions being more attractive to the peptidase than the other. But it also indicates that one would have longer lasting effects than the other. "So I think for many of them, the key, if there is a functional basis for their numbers, may not be at the level of the receptor, but it may be much more at the level of their time course, or their stability in the hemolymph—in which case it could be very physiologically relevant. But you would be missing it if you just did a rapid assay."

Marder has long been convinced that only by identifying all the elements of modulatory control in the stomatogastric ganglion will the general principles be fully understood. Hence, after the novelty of identifying the first modulatory substances, Marder has carried on her peptide hunt in the face of some quite disparaging comments. "After we did the first modulators, people kept saying: 'Why are you going after another one? It's just another modulator, nothing new.' I kept on saying we're not going to understand how the circuit works until we get them all, and understand them all. My goal is to get the secret of why there are so many." Of course, that has meant publishing quite a few papers that Marder calls banal and boring, although they took just as much work as the earlier, innovative ones. It was all part of developing the whole cast of characters. Some of them were intrinsically interesting to study, and others were just adding to the list. "If you wanted to only publish in *Neuron*, you couldn't do this work. But the payoff is down the road when you have twenty five of them and you can say this network is modulated not by two things but by twenty five things—that's a very different statement that you could never have made if you hadn't been ready to do the tedious work."

A newer method called "MALDI imaging" now makes it possible to run MALDI on an anatomical structure. The laser moves across a section of tissue, with its structure intact, and gives a spectrum at every point. As the resolution of this technique improves, it should be able to show which ones of those peptide family members are found where, thus answering two more unresolved questions: Are different versions of the same peptide differentially expressed in different parts of the same neuron coming into the ganglion? Do different neurons express different members of a peptide family? That sort of study might shed light on the physiological significance of the different subsets of isoforms, indicating whether they are released into the circulating fluid under different conditions, especially at different stages in the life cycle or under stress. Then there is the extremely complicated question of the relative quantities of different peptides that are present in the ganglion at any one time. Most stomatogastric ganglion physiology has been done one peptide at a time; the peptide is applied, and the resulting electrical activity is measured. But in the natural environment, in living animals, the stomatogastric ganglion is bathed in a complex and constantly changing chemical soup made up of hemolymph and a combination of substances released into it. Lingjun Li's lab has developed methods to study these changing concentrations under different physiological conditions, but the multiple effects of neuromodulation will not be completely understood soon, even in the stomatogastric ganglion with just thirty neurons.

When Marder started her research, neuroscientists knew that substances were released from nerve cells, but they had no understanding about how they worked. In the first years in her own lab, Marder learned a great deal about the curious neuron with all its tricks and subtleties. Because so little was known at the beginning of her career, and because techniques were limited, her experimental work in that period may seem laborious and haphazard. But I think that's relevant to her development as a scientist and to examining her thought processes. Precisely because she was in the dark with a weak flashlight, she had to make the most of anything she could see, any data she could get. Even if she had a hypothesis, often the techniques did not yet exist to examine it. So she learned to use her analytical and creative mind to wrest significant meaning from all the clues she could get. By rigorous forensic reasoning, with her growing understanding of the links between an animal's physiology and its behavior, she was able to integrate seemingly unrelated bits of data into fundamental functional principles.

6 The Multifunctional Network

When Eve Marder's tenure was confirmed in the spring of 1984, she went to the chair of her department and told him her lab was too small. Autumn 1984 found the new associate professor in a bigger lab, ready to expand her team and her horizons. With her exposition of the concept of neuromodulation, Marder had reached the forefront of her field after only six years in her own lab and was opening up new avenues for her science.

The years of pharmacological investigation and her focus on identifying the neurotransmitters of the stomatogastric ganglion had turned out to be the key that opened the door to understanding neuronal networks. Pharmacology was easily sneered at as just throwing drugs on preparations to see what happened. It was not until the elegant and purposeful single-cell work such as Marder's that people understood that pharmacology could reveal circuit function. Answering her own earlier questions about why so many substances were present in the stomatogastric ganglion, she had shown that the neurons of the ganglion used only two of them, acetylcholine and glutamate, as neurotransmitters. The rest of the ever-growing list of substances circulated in the hemolymph or were secreted by neurons outside the ganglion that had fibers in the incoming nerve reaching its neuropil. The influence of these inputs could modulate the responsiveness of individual neurons in the pyloric circuit. Impressively, she had created a framework for understanding their purpose, taking neuroscience further than her experimental observations alone might suggest. With Judith Eisen, she had shown that such neuromodulation, affecting the individual neurons in different neuron-specific ways, could alter the pyloric circuit's frequency and the phase relations within its cycle, so modulation at the level of the single neuron implied that the network as a whole must be modifiable. Thus, the role of neuromodulation could be to enlarge the repertoire of responses of the whole circuit, not just to affect a lone neuron.

Marder didn't coin the term "neuromodulation," but she was certainly among the first to put forward an explanation of its function, publishing a clear description of her concept in February 1984, whereas a search for "neuromodulation" in the archives of the journal *Nature*, for example, produces the earliest mention in June 1985. Research in other labs was tackling various aspects of modulation, and in 1980, Kravitz and his team had published a description of the effects on motorneurons controlling posture in lobsters and crayfish of injections of serotonin and octopamine into the hemolymph, in vivo, with this conjecture: "Perhaps the postures we observed were only a gross manifestation of what is a very subtle control system." But most researchers of that period were looking for evidence that a substance—any one they had picked as promising—could enhance activity in cells, and to do that they concentrated on a single neuron, a single current, a single type of ion channel at a time. Marder's contemporaries were talking about "control of neuronal excitability," and in looking through archival material, I have found no one working on more than one neuron. Marder, however, was being less reductionist, asking what the functional consequences of changes in neuronal excitability were and investigating the changes in behavior of groups of neurons such as the AB-PD-LP set or the whole stomatogastric ganglion when exposed to influences other than synaptic neurotransmission.

Several lines of research were running in parallel in her lab. The identification of further neuromodulating peptides and the description of their effects on the ganglion continued, but now the goal of understanding the circuit mechanisms by which connected neurons interact became the focal point of her work.

Marder's conjecture was that neuronal and circuit mechanisms had to be flexible to provide for the full range of an animal's behavior with its smooth passage between variations of basic movements. Otherwise, *reductio ad absurdum*, the animal would need a dedicated circuit for each variant movement. She saw that, in a circuit like the pyloric circuit of the stomatogastric ganglion, neurons could be variously modulated by multiple substances, meaning that the circuit could produce multiple outputs. Marder started to refer to "the multifunctional network," and she probably was the originator of the term "circuit neuromodulation."

* * *

One of the most important and most studied tasks of neuronal circuits is to produce the rhythmic repeated movements on which animals depend, such as breathing or swimming, or, in the case of the lobster, moving food in its stomach. How such rhythmical, repeated, motor actions were initiated and controlled was a puzzle that occupied the minds of many biologists. The hypothetical mechanism had been called "the central pattern generator." The hallmark of a central pattern generator is that its rhythm can arise within the network itself—the network does not need to receive any rhythmic or alternating input from some other part of the animal's nervous system.

What kinds of behaviors do central pattern generators control? Some of them are lifelong, continuous movements such as breathing, some are responses to contexts such as walking or chewing, and others are brief, such as the crayfish escape pattern that Allen Selverston's early experiments described—entirely stereotyped responses to danger. These movements may influence each other, so, for example, when you break into a run your breathing speeds up to match your respiratory needs. They have to be flexible: you don't walk like a robot, you adjust your stride to go uphill or over stony ground. So sensory feedback must be involved in the living animal, but a central pattern generator doesn't depend on it; a neuronal pattern-generating circuit can produce its rhythm even when cut away from the rest of the body.

The stomatogastric ganglion is ideally suited to studying central pattern generation. The triphasic motor pattern of its pyloric rhythm may be altered by neuromodulation caused by an animal's state, but it is ongoing throughout the animal's life. Indeed, the isolated ganglion in a dish continues to produce motor patterns—"fictive" ones—much like those of the living animal. In most motor systems, the central pattern generator mechanism originates among interneurons, and these drive the motor neurons. But in the stomatogastric ganglion, the motor neurons contribute on equal terms with the few interneurons to generate rhythms. Therefore, the pattern generated can be recorded from the motor nerves, which makes the ganglion relatively easy to work on. By the mid-1970s, in the light of its connectivity work, Selverston's lab was investigating the ganglion explicitly as a central pattern generator, and Marder was aware of this aspect of his work.

Many neuroscientists were still expecting that these patterns of circuit activity would be explained by special properties of circuit wiring—it was even thought that there might be just one mechanism used in all rhythmic behaviors. They were looking for *the* central pattern generator, the universal, ideal pattern generator that would be applicable to most, if not all, movements. At the least, there would be the same circuits in all animals that have similar motor patterns, stemming from a prototypical central pattern generator whose essential mechanisms would have been conserved throughout evolution. But, as researchers worked out the leech heartbeat system, the locust flight system, and the *Tritonia* escape system, they could see that neuronal circuits that produce similar motor patterns could have different underlying structures. By the early 1980s, it was clear that every circuit was different, idiosyncratic, and peculiar to that species; no general principles had been found. The discouraging perspective of analyzing untold thousands of circuits opened up before their eyes.

Among the researchers engaged in this pursuit was Peter Getting, an exceptional scientist of flair and intuition. His background was in biophysics, and he was a pioneer in using computational modeling to try to understand the dynamics of networks. During the 1982 Society for Neuroscience meeting in Minneapolis, Marder struck up a conversation with him at a party. The hubbub threatened to overwhelm their voices, but it was the only opportunity to talk because Getting was due to fly back to Stanford the next morning. They retreated to a bathroom and spent a couple of hours there talking over their ideas. The long conversation cleared up some loose ends in both their minds, and they undoubtedly influenced each other. Getting was working on one of the sea slugs (*Tritonia*), studying its escape swim in which this unprepossessing animal repeatedly doubles over forward and then arcs back in what are called ventral and dorsal flexions. He had just put forward the notion of a "polymorphic" network. Marder's concept of the "multifunctional" network was close, although they had used different experimental and thought processes on the way. Both had come to the conclusion that networks could be flexible. Getting, says Marder, "had come up with less good evidence but greater insight. He was studying a sensory neuron that had a different impact on the network depending on different parameters. Judith and I had come to a similar conclusion by another route. We had gotten to that point after seeing the plasticity of the phase as a consequence of modulation." Here she is referring to their finding that, depending on their activation relative to each other, the AB and PD neurons could affect the phase relations in the pyloric rhythm.

Getting's response to the failure to find an archetypal central pattern generator was to analyze, in abstract terms, what the properties of circuits generating repeated rhythms would have to be. His work in the early 1980s led him to propose a possible synthesis between the two contemporary hypotheses, which were either that the central pattern generator was a group of neurons that independently drove motor neurons to sequences of coordinated activity with a temporal rhythm or that there was a "command neuron" or group of neurons capable of initiating rhythmic behavior patterns. Getting proposed that a rhythmic motor system could be represented as:

SENSORY INPUT → COMMAND → CPG → MOTOR NEURONS → BEHAVIOR

This separated the command from the pattern generation since the command was activated by sensory input and then triggered the central pattern generator. It left unanswered the question of what might bring the activity to an end and was complicated by the fact that in Getting's molluscan preparation, two of the neurons Getting had identified as command neurons also participated in the generation of the swim pattern. After his discussions with Marder, he began to call these "multifunctional neurons."

If there was no single underlying circuit architecture, then the question might be what primitive elements would be conserved, what functions would contribute to pattern generation? Getting turned to computer modeling. He used data from his experimental preparation in *Tritonia* as input for his models, making specific sets of measurements and then changing parameters in the model one by one to see what happened. He was looking for discrete circuit mechanisms that could be recombined in all sorts of different ways because he had conceived the idea of a components system: subcircuits would be the building blocks that could be put together to form movement-generating circuits of different kinds. He soon acknowledged that the models were actually just telling him what he might already have known but hadn't recognized until he saw the computed results. But they helped him develop a library of functional building blocks, including phenomena such as reciprocal inhibition between neurons, particular properties of each neuron's membrane, and other mechanisms that had recently been shown to contribute to circuit dynamics. He was, Marder says, "unquestionably the most influential small-circuit person around in the mid-1980s. He had a very clear voice, and he had a way of speaking that all the boys understood. He had a clarity of mind that somehow or other got through to them."

Marder's approach was markedly different, and more nuanced, because she had realized that neuromodulation was part of the central pattern generator puzzle. It was now a notion well supported by experimental evidence. She had articulated a theory that framed the observed phenomena, showing that central pattern generators were not explained solely by the connectivity of the neurons and proposing neuromodulation as the explanation of how their rhythmic patterns might change in response to sensory input. But these ideas were novel; she was young and, still pertinent in the

1980s, she was a woman. She puts it like this: "Well, actually, I didn't always see everything exactly the way Pete did, but he had the ability to make them understand. I didn't have it then the way I have it now. Either I've learned how to do it or I've learned not to care what the men say."

Most of her contemporaries thought that neuronal excitability was a fixed property and that all stable change in circuits would be due to changes in synaptic strength, that is, the strength of the connections between neurons. But Marder was telling them that neither of those suppositions was accurate, that the dynamics of the cell and its intrinsic excitability could be changeable and could affect a circuit's output without altering its connectivity. In the review mentioned in the last chapter, appearing in *Trends in Neuroscience* in 1984, Marder examined the stomatogastric ganglion explicitly as a central pattern generator. She brought together the concepts of neuromodulation, the multifunctional circuit, and central pattern generators: "we are now at the stage of beginning to understand how the output of central pattern generators can be altered, thus enabling them to show the full richness of their behavioral repertoire." It was an important publication for Marder, bringing her to the attention of a wider audience of researchers, especially those studying the central pattern generator, and it brought the concept of circuit neuromodulation into their everyday language.

Her ideas, and those of Getting, appeared in 1985 in two chapters of a book edited by Allen Selverston. They were influential. Getting's chapter principally describes the synaptic interactions leading to pattern generation, such as the strength and duration of sensory input. However, he had found Marder's arguments persuasive, and his chapter ends with the suggestion that subtle changes involving modulation might alter the strength of synaptic connections and intrinsic membrane properties. Marder's chapter (written with Scott Hooper) concentrates on the variety of stable motor patterns that different neurotransmitters and neuromodulators could produce, but the last few pages discuss the burning unanswered questions with prescience and confidence. The final summary paragraph is almost a parody of the building block metaphor: "When applied singly, each substance produces a unique constellation of changes in the outputs of the stomatogastric ganglion neuronal circuits. This provides an alphabet of modulatory actions, which then can be combined into a lexicon of behaviorally relevant modulatory changes in the motor pattern."

In 1988, Getting's active life was ended by a disastrous stroke. He had been a leading figure in the world of invertebrate neuroscience, and he left a vacuum in the field. Marder had considered him a real friend; her respect and admiration for him are enduring, and to this day she often cites his seminal work in her major lectures. "His last review article is a really remarkable read. He was a wonderful man. The whole field got dealt a very bad blow when he was knocked out. It took us a while to sort of recapture momentum, probably took until I grew up, fully, because nobody else ..."

She didn't finish this thought, so I prompted: "Was it one of the things that made you grow up?" After a pause, she sighed, "Maybe." What is clear is that, over the following years, the research groups studying invertebrate science picked up speed again, and Marder eventually emerged as one of the foremost voices in the field.

In the end, decades of work have revealed a great variety of basic cellular and biophysical pattern-generating mechanisms; among them, a number of canonical small circuits—building blocks—that are found in numerous systems. All pattern-generating networks are subject to multiple influences—sensory feedback, hormonal influences, and modulatory inputs—the physiological responses to the animal's varying circumstances. But even today, although a number of circuits are well understood, there is no comprehensive description of how the properties of participating neurons cooperate to form a functional circuit or how oscillatory circuits can provide an animal with so many coordinated behaviors. Probably there never will be because of the idiosyncrasies of specific circuits, each with its own evolutionary history.

* * *

Marder's lab had shown that the neuropil of the stomatogastric ganglion included dense and widespread fibers containing proctolin. These came from neurons with cell bodies in other ganglia. Scott Hooper's work had demonstrated proctolin's strong modulatory impact on the dynamics of the pyloric circuit. He had shown that bath applications of the peptide onto a silent ganglion started up its rhythm or speeded up a slowly bursting preparation, whereas it had little effect on a strongly running one. Now he came up with a bold and original idea: he would try to account for all of proctolin's effects on the circuit on the basis of its effects on the individual neurons: which cells were the target for proctolin and what did it do to them? He would have to isolate each neuron in the circuit and study the effects of proctolin on it. Systematically, using both pharmacology and cell kills, he set out to identify the neurons that responded directly to proctolin and to distinguish them from the neurons that did not but were affected "secondhand" by responding cells.

It was a brilliant piece of work. He discovered, first, that only the AB (anterior burster), LP (lateral pyloric) and PY (pyloric) neurons responded to proctolin, implying that they had receptors for proctolin in their membranes, whereas the VD (ventricular dilator) and the two PD (pyloric dilator) cells presumably had none because, when isolated, they were unaffected by it (see figure 6.1).

Second, to everyone's surprise, including Marder's, he found that however slowly the intact pyloric circuit was running at the beginning of an experiment, in proctolin, it would quickly settle at 1 Hz (1 cycle per second, 60 cycles per minute), which is its characteristic frequency. But that was not because the pacemaker AB cell went to 1 Hz in proctolin; rather, when isolated, the AB cell contrarily went faster than the circuit, to more like 2Hz. That stumped them because they expected the fastest oscillator to drive the frequency of the network, as it does in the physiologist's standard model, the heart, where the fastest of the pacemaker tissues, the sinoatrial node, sets the pace. But here the oscillator was going faster in isolation and more slowly in an intact circuit. Because the experiments showed that proctolin

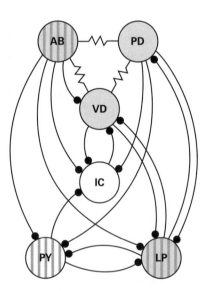

Figure 6.1
The neurons of the pyloric circuit showing those affected by proctolin. Grey tone indicates neurons that show altered activity when proctolin is applied to the intact circuit. The AB and LP neurons show activity changes in the presence of proctolin and, indicated by vertical lines, are also affected by proctolin even when isolated from the circuit. Circles indicate inhibitory chemical synapses. Resistor symbols indicate electrical coupling. Neurons: AB, anterior burster; IC, inferior cardiac; LP, lateral pyloric; PD, pyloric dilator; PY, pyloric; VD, ventricular dilator. (After Hooper and Marder, 1987.)

had no effect on the PD cells or the single VD cell, Hooper posited that the AB's electrical coupling with them was somehow acting as a brake.

Hooper devised another efficient experiment to find out. Using just the electrically coupled cells–the AB, two PDs and the VD–he put them in proctolin, recorded the effect, and then killed off the cells sequentially. The result was striking: as he went along, first killing one PD cell, then the other, and finally the VD, the frequency increased. This observation confirmed that the AB cell was being slowed down by the nonresponse of the cells to which it was electrically coupled, and could be explained by considering that, through the electrical coupling, the surface area of the PD and VD nonresponsive membrane was, as it were, added to that of the AB, making the ensemble less responsive. The combination, intriguingly, ensured that the circuit went to its normal frequency.

Their paper describing the work, published in 1987, is one of Marder's favorites, and it's easy to see why; it has an energy and a confidence that it is opening up a new way of looking at circuits that makes it an exciting read. It was a ground-breaking experiment, the first example of teasing out the individual cellular responses that contributed to circuit change caused by neuromodulation. It also increased the scope of the neuromodulation phenomenon, showing that the pacemaker was not the only controller of the circuit's frequency: a modulatory input could control the controller.

The discussion section of the paper is, of course, based on careful and detailed analysis of the experimental results, but it bristles with suggestions and speculations. Hooper and Marder presented the contradiction that what they called "one of the implicit assumptions of neuropharmacology"—the importance of knowing which neurons have receptors and can respond to a substance—was partially negated by the fact that "neurons that do not have proctolin receptors shape the physiological responses of neurons that do." This meant that it was insufficient to rely on receptor and transmitter distribution studies as indications of circuit behaviors. They suggested that some of the connectivity and some of the membrane properties of circuit neurons might be necessary, not for the generation of the basic rhythms, but only to provide sensitivity to different inputs to the circuit. That would enable a wider range of muscle control and behaviors in response to the ganglion's environment and the events in the animal's life, such as swallowing food. Finally, they explicitly proposed that the function of neuromodulation was to influence the output of an entire neuronal network rather than the individual neuron members of it. Obviously, this supported Marder's concept of the multifunctional network.

Marder and Hooper thought they had found a truly general phenomenon that would be applicable to all electrically coupled ensembles of neurons, but it turned out to be only half the story. Marder would discover the other half, with amazement, in later theoretical work (and in the next chapter). Hooper then left the lab, going to France to start a postdoc with Maurice Moulins, one of the patriarchs of the stomatogastric ganglion field. Like Allen Selverston, Moulins had learned the stomatogastric system from Don Maynard.

Marder says she ended up learning an important lesson from Hooper. "At a certain point in Scott's project, I realized that he had framed a question that I hadn't thought of. That was really interesting and important. My first experience of saying, oh, this project has gone way beyond what I thought it was. Judith and I moved together, but Scott saw something that I hadn't even envisioned. So I learned from that: the smart advisor encourages the creativity and intelligence and drive of their people and benefits from them. I also realized that you know when someone's ready to graduate. When they have taken deep ownership of their project so sometimes you're learning from them. Or they've seen or understood something you haven't. And then you know it's time for them to graduate."

Marder's parting words to Hooper as he walked out of her office were, "Follow the data. It will tell you what to do."

* * *

So far we have mostly been concerned with the pyloric rhythm rather than that of the gastric mill. Allen Selverston and Brian Mulloney had mapped the gastric circuit's connectivity while Marder was a doctoral student in San Diego. To researchers, the gastric mill rhythm seemed less reliable than the pyloric circuit that chugs along helpfully in the dish. Out in the rocky shallows, the pyloric rhythm constantly ticks over at some level of activity throughout the animal's life, whereas the gastric circuit is active only when the animal has been feeding. Also, even when it is active, the gastric rhythm is slower than the pyloric rhythm, although it can generate emphatically strong rhythms. It has no clearly identifiable pacemaker neurons.

In the mid-1980s, working in Selverston's lab, Georg Heinzel used an endoscope in the stomach of a lobster to film the movements of the gastric mill teeth. He showed that there are two variations on the gastric theme: the gastric muscles operate both a squeeze mode and a cut-and-grind mode. Marder reviewed the work for *Nature* in 1988, rightly calling it "a technical *tour de force*." It was also unforgettable. At a Society for Neuroscience meeting, Heinzel presented his film to an ecstatic audience so appreciative of its suggestive images that they gave the surprised young scientist a standing ovation. A glance at Plate 4 will enlighten you.

In 1986, one of Marder's graduate students, James Weimann, decided to study the gastric mill rhythm in the crab. Despite many careful record-ings from the gastric circuit neurons, he couldn't pick up a gastric rhythm at all. Eventually, he went into Marder's office, obviously ruffled, and said he had recorded from every single cell in the ganglion, and they all fired in pyloric time. "There *are* no gastric neurons in crab, they're all pyloric." Marder calmed him down and pointed out that there were certainly gastric muscles, even in the crab, and they were driven by neurons. She sent him away to record from those muscles and the neurons together.

Surprisingly, Weimann discovered he had been right, in a way. The gas-tric circuit neurons did exist, of course, but all except one were misleadingly firing in pyloric time. As long as the gastric rhythm wasn't called for, they just switched to pyloric time. It was only when he treated the cells with a certain peptide (named by its structure, SDRNFLRFamide) that he could pick up the gastric rhythm. This work revealed for the first time that, under certain modulatory conditions, neurons of one circuit could join in with another circuit's rhythm (see figure 6.2).

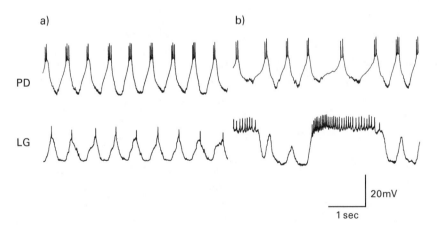

Figure 6.2
Traces from two motor neurons considered to belong to different circuits: the pyloric dilator (PD) and the lateral gastric (LG). (a) Left-hand side, when there is no rhythmic gastric mill activity, the LG neuron fires in phase with the PD cell at pyloric circuit frequency of about 1 cycle per second. (b) Right-hand side, during gastric activity, the LG neuron fires strong bursts at a lower gastric rhythm frequency. The two condi-tions were recorded from the same cells. (After Weimann et al., 1991.)

Weimann was also able to show that at least two of the neurons capable of switching their allegiance between the pyloric and gastric teams could actively influence both circuits. These neurons were not just following some strong drive; they took on a functional role in either circuit, and if one of these double agents was perturbed experimentally, it could reset the circuit's timing. Thus, it could be said that individual neurons were full participants in more than one pattern-generating circuit. Further, Jim Weimann and Pierre Meyrand showed that the converse also occurred: in the absence of any stimulus, some of the neurons traditionally described as pyloric would spontaneously fire in time to the gastric rhythm. This switching was news to neuroscience, and the paper reporting it was influential.

In the crab, some of the stomatogastric ganglion motor neurons control both muscles of the pyloric chamber and muscles that move the gastric mill teeth. It seemed logical that these neurons should respond to both pyloric and gastric rhythms. Heinzel then showed that the lateral teeth in the gastric mill did indeed move in time to the pyloric rhythm when they were not strongly activated to chew or grind. These endoscopy results were most valuable, relating the electrophysiology to observed muscle movement, demonstrating that the in vitro results properly reflected natural behaviors.

All these lines of evidence added up to a firm result—in the laboratory. But Marder had her usual nagging doubts on a further point. Did the experimental pharmacology reflect reality? For example, in the living animal, did the stomatogastric ganglion really receive any proctolin input and react to it? Quite a lot of evidence hinted that it did. Among at least a dozen corroborating studies, I'll mention just two. The immunohistochemistry described in the last chapter showed the dense and widespread presence in the neuropil of proctolin-containing fibers originating from neurons in other ganglia. Further in vivo endoscopic work by Heinzel showed that when proctolin was injected into the hemolymph, it activated muscle movement that matched the activation of motor neurons by proctolin in bath applications to the ganglion in vitro. It looked as though proctolin could safely be thought of as a naturally occurring modulator that influenced digestive behaviors in the living animal.

That left the question of how proctolin and other neuromodulators reached the ganglion's neurons. It was thought that the neurons of the esophageal and commissural ganglia were the source of most modulatory substances. It was always obvious that preparations of the ganglion that included the front end were the most reliably active, suggesting that activating modulatory inputs probably came from these ganglia. But were the modulators secreted at synapses within the neuropil? Or were they secreted as "volume transmission," which means diffused secretion from the axon of a neuron into extracellular space? Or were they perhaps secreted into the hemolymph by even more distant neurons? How could you show that the experimental application of, say, proctolin, to the stomatogastric ganglion was an acceptable imitation of natural processes?

A new postdoc in Marder's lab embarked on the most rigorous effort yet to relate the observed actions on the ganglion of possible neuromodulatory substances to the secretion or neuronal release of these substances outside the ganglion. Mike Nusbaum extended the lab's investigations to the nerve fibers descending to the ganglion; he wanted to stimulate the nerves that were thought to secrete a particular peptide, rather than dump that peptide onto the ganglion. The lab's immunocytochemistry had shown that there were cells in the esophageal ganglion that stained for proctolin. Nusbaum showed that these neurons sent axons in the stomatogastric nerve to the neuropil of the stomatogastric ganglion. Then he stimulated the proctolin-containing neurons and recorded initiation of pyloric circuit activation that was similar to what Hooper had described. He published the work with Marder in 1989, and these neurons acquired the name "modulatory proctolin neurons" (MPNs, which just happen to be Nusbaum's initials). To Marder, "That was an important step. Mikey has always been the advocate for considering the information you get from bath application of these modulators very different from what you get when you use the neurons themselves, for all sorts of reasons." Since then, from his own lab, Nusbaum has demonstrated the exactness of his conviction; different stomatogastric nerve neurons using the same peptide transmitter turn out to have different actions on network activity.

Marder calls the late 1980s a golden era in her lab. "That was the heyday of breaking open neuromodulation when Scott had done this beautiful story with proctolin and Mikey did the first studies of the new era of the descending projections. Just when Scott left, Patsy Dickinson was here on sabbatical, and Mikey, Pierre Meyrand, Jim Weimann, and Jorge Golowasch were all here. It was a really quite wonderful time."

Nusbaum says, "She was there all the time, working late at night and going around to everybody to see what they were doing, with ideas and suggestions. I was a starting postdoc—I was closer to just generating data than putting them into contexts that were more than one step away from the data. Marder is remarkably creative. She can take some successful stuff that someone who wasn't such a broad thinker might have packaged in a pretty narrow context and put it into a much wider context without falsely enhancing it in any way. Or she'll suggest just a little bit more work to make your results a lot more profound."

Their own excellent work notwithstanding, Marder's collaborators always acknowledge her long hours of work, and talk about her profound insight and intellectual leadership with awe. Soon, however, Marder had to step back from work at the bench. The lab got bigger and her days busier. She stopped working at the rig first because it required long stretches of sequential work, and she had to come and go on other business. Even if she could find time for it, she says, "What killed me was moving from using tape recorders and chart recorders to computers. We kept on changing the acquisition and analysis software. Every time that software is upgraded, the students and the postdocs doing it day by day, they learn the new system, they upgrade themselves. But you can't put it down and go back six months later and have the foggiest idea how those programs work."

For many years, she did a lot of the immunohistochemistry herself or shared it with Nusbaum. She took the pictures on the microscope and did all the darkroom work for many of the publications. Here, too, the end came with the confocal microscope and the constantly changing software packages that run it. But the lure of hands-on work was still there, and she would be in the lab trouble shooting whenever a new technique was on trial, from preparing samples for MALDI in the mid-1990s to doing some of the lab's first polymerase chain reactions (PCRs), a technique of molecular biology, in 2006, because she wanted to know how it was done.

* * *

The next unpredicted phenomenon was discovered in serendipitous fashion by Patsy Dickinson, an energetic young associate professor on sabbatical leave from Bowdoin. Serendipity features often in Dickinson's story. To begin with, she was working on the stomatogastric ganglion only because her PhD advisor forgot to answer a letter from a French colleague offering a postdoc position. He persuaded Dickinson to save his face by taking the job. She thought, "Even if it's not a good model, I'll learn French. So I went to Arcachon to work with Maurice Moulins. It turned out to be spectacular, and I've been working on the stomatogastric ganglion ever since." Back in the United States, she went to the Society for Neuroscience meeting in 1982. She was presenting a poster about a modulatory neuron in the stomatogastric nervous system. Marder walked up to the poster and exclaimed, "That's it! That's that cell!" They started an animated discussion that still continues whenever they meet.

Dickinson worked on *Panulirus interruptus* in her own lab and was interested in the cardiac sac, the initial store for ingested food. No one had paid much attention to the cardiac sac rhythms in the spiny lobster, for good reason. Although its two controlling motor neurons were unambiguously identified, their cell bodies are in two different ganglia, the esophageal ganglion and the stomatogastric ganglion, and they both project axons to the commissural ganglia. Because three ganglia are involved, it's not easy to devise experiments that can separate out the effects of different interventions. Moreover, the cardiac rhythm is frequently silent in vitro, and when it does run, it has a long natural cycle period that can be measured in minutes, even slower than the gastric rhythm, and much slower than the pyloric cycle of one second. What is worse, that cycle is irregular, making it harder to establish a baseline so that change can be quantified (see figure 3.1 in chapter 3).

In Marder's lab, Mike Nusbaum had been studying a peptide called Red Pigment Concentrating Hormone (RPCH) that affected the pyloric rhythm in the crab and also strongly activated the two motor neurons dilating the cardiac sac. So, in her sabbatical year of 1987–1988, Dickinson decided to look at the effects of RPCH on the cardiac sac rhythm in the spiny lobster.

Her first fortuitous discovery in that rich year followed a face-off with Marder about whether proctolin, too, would activate the cardiac sac rhythms. Marder was thinking of Hooper's work, which had suggested interactions in the cardiac sac, and was sure it must have an effect. She knew the peptide was found in all the ganglia of the stomatogastric nervous system, and so she assumed it would affect some of the neurons in each of them. But Dickinson couldn't find any evidence of it. "I told her, absolutely not, I've put it on a million times, and it just doesn't do that."

One afternoon they were sitting around arguing about it until Dickinson dragged Marder to her rig to show her. "I put on the proctolin, but I hadn't washed out the RPCH for nearly as long as I usually did. I had the whole stomatogastric nervous system in the dish, but I put the proctolin just on the stomatogastric ganglion, and these perfect cardiac sac rhythms appeared! Just totally random! Because I happened to have done it without a thorough wash."

Marder's assumption was correct; it was just a question of finding the right conditions. It turned out that, for proctolin to have this effect, RPCH had to be present, even in low concentrations. This wouldn't happen often in the laboratory because most experiments analyze the results of applying one specific modulatory substance at a time, but of course in the living animal, the mix of chemicals in the nervous system's environment supplies multiple influences from different sources all acting simultaneously.

RPCH, Dickinson found, could activate the cardiac sac motor pattern wherever it was applied, and all parts of the stomatogastric nervous system seemed to be affected by it, albeit to different degrees. She also observed that applying RPCH to the stomatogastric ganglion alone activated the cardiac dilator neuron (CD2) within it. Most of the time, except when vigorous movements of the sac are called for, the CD2 runs along slowly. In RPCH, it started bursting and the anterior median neuron (AM) started firing in alternation with it. The AM controls the constrictor muscles of the cardiac sac, but typically takes its cues from the gastric mill rhythm. Dickinson was analyzing these different effects in more detail when she stumbled on something significant.

One day Marder happened to go by her rig just to see what was up. Dickinson was recording her control traces for an experiment. Then she put RPCH onto the nerve fiber between the esophageal and stomatogastric ganglia and immediately started turning down the gain on the chart recorder. Marder said, "Patsy, what are you doing—why are you turning down the gain?"

Dickinson replied, "Well, if I don't turn down the gain, the postsynaptic potentials get so big they go off the scale."

"Wait—you mean RPCH increases the sizes of these PSPs so much that they go off scale?"

"Yeah," said Dickinson, "I always have to turn down the gain."

A second later, Marder said, "Well, du-uh!" and Dickinson shrieked, "Oh, oh! Of course!"

And that is how they discovered that RPCH could so strongly amplify the synapses made by cardiac sac neurons onto neurons in the stomatogastric ganglion that the cardiac rhythm drew the gastric circuit neurons toward its own cycle. A new rhythm took shape, neither one nor the other: it was much faster than the cardiac frequency but slower than the gastric, and it didn't really resemble either (see figure 6.3).

Marder and Dickinson called the phenomenon the "fusion" of rhythms. It was quite different from the switching of individual neurons between two different ongoing rhythmic motor patterns. In the case of fusion, two independent rhythms of different frequency, governing independent sets of muscles, converged into a new rhythm unlike either of the original two, which meant that a different output could be produced. They published their report with the title "Neuropeptide Fusion of Two Motor-Pattern Generator Circuits," and it appeared in *Nature* in 1990. The work demonstrated a powerful effect of neuromodulation that took its impact well beyond just

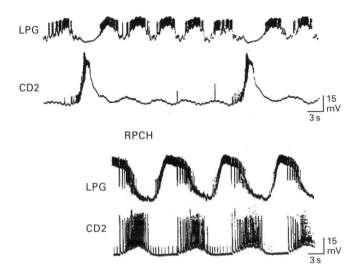

Figure 6.3

Electrophysiological traces of cardiac, gastric, and fused rhythms. All these recordings were made in the same preparation of an intact stomatogastric ganglion. Upper
panel: The lateral posterior gastric neuron (LPG) fires in gastric rhythm with a period
of 5 to 10 seconds. The cardiac dilator 2 neuron (CD2), recorded simultaneously with
the LPG, has a much longer period, sometimes measured in minutes. The period of
the neuron recorded here is at the short end of the range. Lower panel: When red
pigment concentrating hormone (RPCH) flowed over the ganglion, the same two
cells fired rhythmical bursts with a new period that is significantly longer than the
gastric rhythm period and significantly shorter than the cardiac. (After Dickinson
et al., 1990.)

raising or lowering a threshold. Neuromodulation could bring about a significant reconfiguring of a system's networks. Marder's schema of neuromodulation and her characterization of the circuit as multifunctional were
now well established.

* * *

Peptides such as proctolin and RPCH, it seemed, could be modulators of
the whole chain of feeding and digestive behaviors. It makes sense that the
cardiac sac movements should be linked to those of the gastric mill; food
has to be moved from the cardiac sac at the right time and in the right
quantities to be masticated in the gastric mill. Evidence for this was soon
available. Georg Heinzel had gone back to start his own lab in Germany,

and Jim Weimann went to work with him for six weeks. They were both stars of technique in the lab, inventive and able to craft just about any piece of equipment. They were also perfectionists. "You have no idea what they did. Two incredibly talented, creative guys," Marder relishes the story. "But neither of them believed in ever repeating anything. In their defense, they were recording with the endoscope. These were hard, hard experiments and getting an electrode into the ganglion—they couldn't exactly plan what neuron they were going to get. They had a ton of fabulous data—done once. And there was probably enough data for eight half papers but I couldn't stitch it together. I was tearing my hair out to make just one paper of all that cool stuff." That one paper came out in 1993 and showed that the activities of the different regions of the stomach were closely linked and their movements could be controlled by several motor neurons that might be considered to belong in more than one circuit.

All this work revealed a most surprising degree of flexibility at circuit level and raised an interesting question: was the accepted distinction between pyloric and gastric circuits valid or even useful? Should the whole stomatogastric ganglion, all thirty neurons, be considered as a single neuronal network, both anatomically and functionally, that could produce several patterns of activity depending on its modulatory state? Or would it be more fruitful to think of individual neurons switching between discrete ongoing rhythms? Marder now thinks a lot of psychic energy was wasted asking questions about circuit boundaries: it is useful to have a practical definition of a circuit to understand how it works on its own, but ultimately the whole nervous system is connected.

I'm reminded of her question as a student in 1971, wondering whether the rhythmical output of the stomatogastric ganglion might be modified by feeding the animal on different schedules, or with different kinds of food, or not feeding it at all. Ten years later, Marder was able to write with conviction: "Animals are remarkably adept at altering their movements in response to changing internal and external environmental demands. The circuits that control respiration, circulation, feeding, and locomotion must be sufficiently robust so that these critical functions are not easily disrupted. At the same time, the circuits must be flexible so that these functions can be modified in response to changing needs."

Thus, by the early 1990s, Marder had led the way to new thinking about pattern-generating circuits. Neurons could switch, under certain modulatory conditions, from being part of one pattern-generating circuit to joining another. A given circuit could produce multiple forms of behavior. A circuit that was functionally distinct under some conditions could be fused with another circuit under other conditions.

Now Marder was asking another profound question stemming from this flexibility: how is the range of modulation reined in and how is "over-modulation" avoided? How do you get predictable and stable behaviors out of circuits that respond so variously to so many influences? She thought that there would have to be limits to the range of flexibilities available to each neuron and each circuit, otherwise the animal would be driven to extremes beyond physiological tolerances. Circuits might be flexible, but there would be rules and constraints that governed them, and discovering these would be an important challenge.

One of those safeguards seemed to be modulation of the modulator. In 1991, Mike Nusbaum, working with Jim Weimann and another graduate student, Jorge Golowasch, made the first electrophysiological recordings from a descending modulatory axon close to its entry into the neuropil. They were able to identify neurons from the stomatogastric ganglion that synapsed onto the incoming axon and influenced its firing. The activity of the incoming stomatogastric nerve axon (SNAX), and therefore the release of its modulatory transmitter, was thus time-locked to the rhythms of the stomatogastric ganglion. The presence of this sort of feedback mechanism highlighted the interdependence of parts of the nervous system and pre-saged yet more complexities to be discovered.

At this stage, Marder was confronted with the pressing need to analyze complex data. Although the wider field of neuroscience tended to think the stomatogastric ganglion was straightforward, at least relatively, now that dozens of modulatory inputs with diverse effects were being identified, describing its circuits looked anything but simple. Circuits would have to be studied under the many modulatory influences, at the level of the individual neuron, at the level of synapses between neurons, and at the level of the whole network and its output.

In thinking over Scott Hooper's intriguing results, Marder regretted the lack of a quantitative analytical method; she could see the limits of qualitative explanations alone. Experimental work at the rig just wasn't going to give her the big picture. To understand the rules governing the behavior of circuits, she would have to deal with the fuzzy realm of interactions between the components and the behavior, where all of the influences and their various effects on different neurons in the circuit meld to produce behavior. These were not linear problems with linear solutions. If she wanted to describe and predict the behavior of a circuit in detail, then she would have to bridge the frustrating gap between the cellular and circuit levels that prevented her from understanding the system. She was turning over in her mind the question of computer modeling.

A nonlinear system's behavior is described by a set of nonlinear equations, and its output is not proportional to the input. Like most biological systems, neuronal networks are nonlinear, and they do not necessarily follow a sequence as one might be led to expect by the frequent use of the word "pathway" in neuroscience. All components can be active, responding and participating at the same time. However, although networks may be unpredictable and can seem chaotic, they are most definitely not random. Animals don't survive by means of randomly operating systems.

Faced with a prodigious multiplicity of variables, the human brain reaches its limits. Only computer modeling would be able to analyze the myriad strands of data to extract their meaning. Marder had no coherent goal as yet, but she was reaching for a level of understanding that would take her further than her "word models" ever could in interpreting complex experimental observations: "a certain kind of crisper articulation that required something formal."

Plate 1

An electrophysiologist's work station. The Petri dish containing the experimental preparation is in the rig, under the microscope and surrounded by modern manipulators. The rack is between the rig and a computer table. (Author's photo.)

Plate 2
Eve Marder at work in front of her rack in Allen Selverston's laboratory at University of California San Diego. Undated, between 1971 and 1974. (Photo courtesy of Eve Marder.)

Plate 3

The Pacific spiny lobster, *Panulirus interruptus*. (Photo: Phillip Colla / Oceanlight.com.)

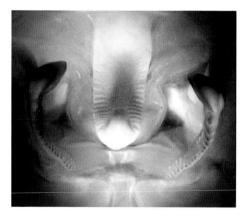

Plate 4

The medial and two lateral "teeth" of the gastric mill in the stomach of *Panulirus interruptus*. Photographed in the living animal by endoscopy. (From Eve Marder and Dirk Bucher, 2007, "Understanding Circuit Dynamics Using the Stomatogastric Nervous System of Lobsters and Crabs," *Ann Rev Physiol*.)

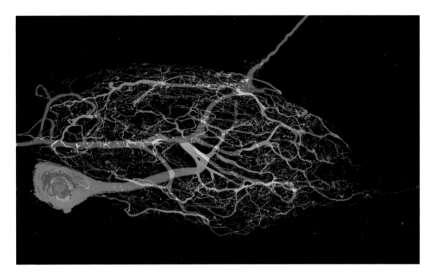

Plate 5

Dye-filled stomatogastric ganglion neurons in *Cancer borealis*. The ventricular dilator (VD) is in green. (Credit: Marie Goeritz, personal communication.)

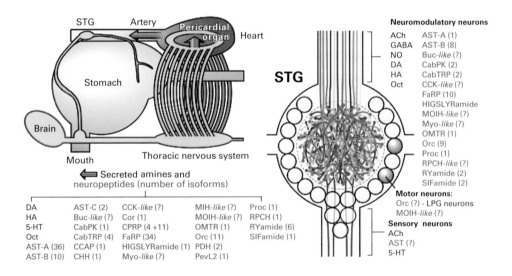

Plate 6

Neuropeptides and amines found in the stomatogastric ganglion. (Courtesy of Dr. Lingjun Li.)

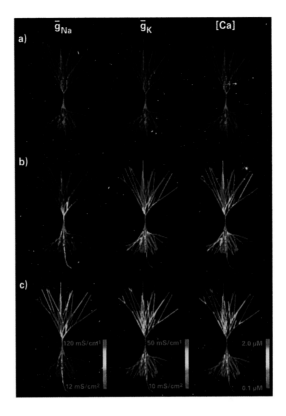

Plate 7

From the left, the three columns show the distribution of maximal sodium (Na$^+$) conductances gNa, maximal potassium (K$^+$) conductances gK, and the average intracellular CA^{2+} concentration [Ca]. Row (a) shows the starting state for this model neuron, with a uniform CA^{2+} conductance but no Na+ or K+ conductances. Row (b) shows the distribution of steady-state conductance after sustained stimulations randomly placed on the lower and upper dendrites. (These would be called "basal" and "apical" dendrites in a biological pyramidal neuron.) The maximal Na+ conductances are concentrated at the soma, whereas the K+ conductances are high along the neurites (conductance strength is indicated as red for highest and blue for lowest). Row (c) shows the result of giving the same sustained stimulation to the lower dendrites, whereas the upper dendrites received much weaker stimulation and developed lower conductances (representing lower numbers of open ion channels in the membrane of a biological neuron). The pattern of distribution of Ca^{2+} concentrations, remaining low in the soma and rising toward the further dendritic zones, had been observed in experiments in real hippocampal neurons. (From Siegel et al., 1994.)

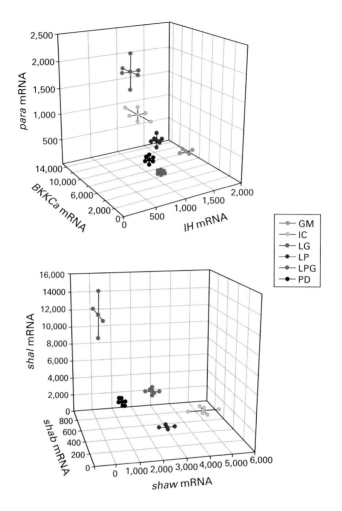

Plate 8

For six types of neurons in the stomatogastric ganglion, three-dimensional plots of expressions levels of six ion channel gene mRNAs show that each type of neuron has a distinct combination; there is no overlap. Neuron types: GM, gastric mill; IC, inferior cardiac; LG, lateral gastric; LP, lateral pyloric; LPG, lateral posterior gastric; PD, pyloric dilator. The mRNAs encode subunits of the protein complexes forming ion channels as follows: $paral_{Na}$—the fast sodium ion conductance, I_{Na}; $BK\text{-}KC_a$—the calcium activated potassium ion conductance $I_{K(Ca)}$; iH—the hyperpolarization-activated inward current conductance I_H; $Shal$—the voltage-gated channel of the transient A-type potassium ion conductance, I_A; $Shab$—I_{Kd}, a delayed rectifier potassium ion conductance; $Shaw$—a second delayed rectifier potassium ion conductance. (From Schulz et al., 2007.)

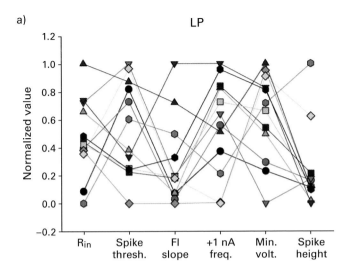

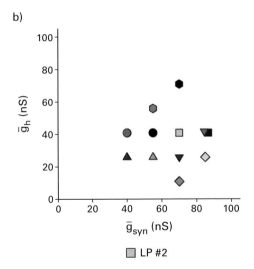

Plate 9

(a) The intrinsic properties measured in lateral pyloric neurons before they were linked into a hybrid network with the model neuron. Each neuron is represented by a different colored symbol, with lines joining the points for that cell. The variability between these cells of the same type is evident. (b) The plot shows a data point for each of 12 lateral pyloric neurons that produced half-center network activity in the hybrid model network. It shows that each cell's particular set of intrinsic properties required a particular combination of input from the dynamic clamp for half-center network activity. Compensation for variable properties is itself variable. (From Grashow et al., 2010.)

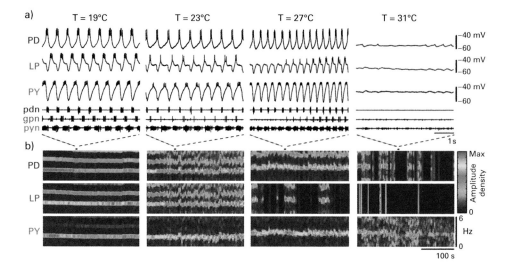

Plate 10

(a) In a preparation from a crab acclimated at 7°C, simultaneous intracellular recordings from PD, LP and PY neurons and extracellular recordings from the pdn, gpn, and pyn nerves that carry their axons show the breakdown of the pyloric rhythm at acute high temperatures. At 23°C signs of disturbance can be seen; at 27 °C there are interruptions to the rhythm, and at 31°C the preparation has crashed. (b) Spectogram representation with dashed lines indicating the timepoints matching the traces above. (From Tang et al., 2012.)

Plate 11

(a) The connectivity diagram of the stomatogastric ganglion of the crab, *Cancer borealis*, showing the five neurons selected: on the fast pyloric side, in red, the PD and LP neurons; on the slow gastric side, in blue, the LG and Int1 neurons, and in black, the IC neuron, which has pyloric and gastric circuit connections. Electrical coupling is indicated by a resistor symbol. Inhibitory chemical synapses are indicated by black circles. (b) The computational model based on the five cells highlighted above. Two fast cells, f1 and f2, are connected to each other and to hn, the hub neuron. Two slow cells, s1 and s2, are connected in the same way. Symbols as above. (c) Voltage traces of components of the model: the hub neuron with an intrinsic oscillation frequency of 0.57 Hz; the paired fast and slow cells with reciprocal inhibition forming half-center oscillators with frequencies of 0.79 Hz and 0.36 Hz; the hub neuron with electrical coupling to one fast and one slow cell, producing synchrony. (d) Examples of voltage traces showing that, with different synaptic parameters, the hub neuron oscillates with the fast cells or with the slow cells. (From Gutierrez et al., 2013.)

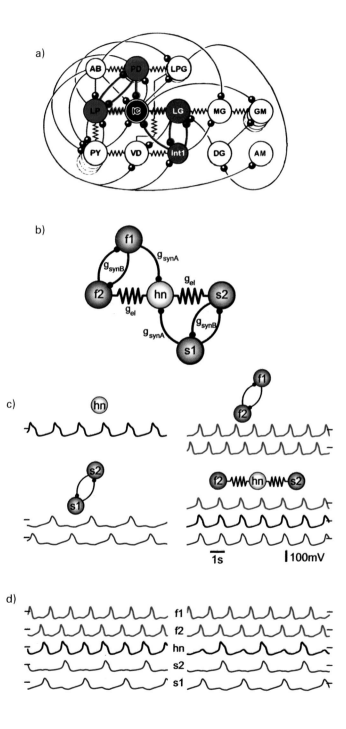

a)

b)

c)

d)

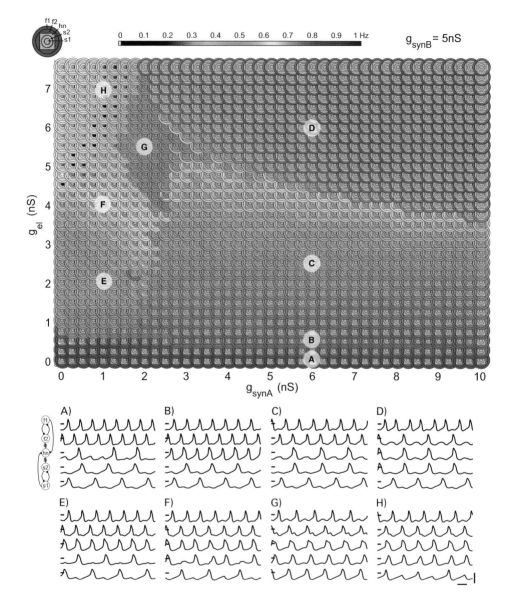

Plate 12

The parameterscape immediately reveals a great deal of information from many relationships in the underlying data. It shows the activity pattern of the model produced at different values for electrical conductance (g_{el}) and strength of the inhibitory synaptic input (g_{syn}). The strength of the reciprocal synaptic inhibition between the pairs of fast or slow cells, g_{synB}, is held at 5 nanosiemens (nS). The model's activity is represented by concentric rings for each of the two fast cells (f1, f2) around a square for the hub neuron (hn), within which smaller circles denote the two slow cells (s1, s2). A color scale from 0 to 1 Hz indicates the firing frequency of each cell at each data point. Different patterns of activity show up clearly, as do discontinuities between them. Voltage traces illustrate the model cells' frequencies at points A to H (from top to bottom: f1, f2, hn, s1, s2). Note, for example, the difference between hn's activity at A and at B; with increased electrical conductance, hn joins the rhythm of the fast cells. But as soon as electrical conductance is increased again, hn rejoins the slow rhythm. With further increases in electrical conductance, hn continues to oscillate at lower frequency, but in region D, one of the fast cells, f2, joins the slow rhythm. The strength of the inhibitory synaptic input is the same for points A to D. Points E to H show the model's responses to lower inhibitory input; in region G, all five model cells oscillate at the same frequency. All traces: vertical scale bar represents 100 mV, horizontal scalebar 1 s; intersecting marks for each trace, 0 mV. (From Gutierrez et al., 2013.)

7 Asking the Right Question

Let's say you start with thirty neurons and a diagram of their synaptic connections to each other. The cell membrane of each neuron is studded with receptors and ion channels in uncounted numbers and varieties. Each neuron uses one of two neurotransmitters, and all respond to dozens of neuromodulatory substances that you know about and certainly many more that you don't. You are lost in a dizzying number of possible interactions.

With electrophysiology, it is relatively easy to uncover the effects of a single neuromodulator on a specific type of neuron. You can also track the dynamics of a small circuit of neurons responding to an environment that includes multiple neuromodulators, but only by its output as a whole. Although you can measure the activity of each one of its member neurons, simultaneously, as they react, simultaneously, to these modulators, you cannot find out why their activity is changing or what the underlying parameters are. What Marder wanted to know could not be revealed by electrophysiology alone. Experimental work would provide more and more detailed data, but she knew it wouldn't help her to understand the big picture, for three reasons. First, no experimental technique was available for monitoring so many variables at once. Second, her experimental data were "idiosyncratic," peculiar to the crustacean stomatogastric ganglion. Third, which is where we started, there were simply too many variables for any human mind to sort out. It was clear that the flexible behaviors of circuits in the stomatogastric ganglion depended on many nonlinear processes simultaneously. How did the multiple, constantly changing properties at the cellular level add up to the changes at the circuit level? As Marder says, the human mind has difficulty "seeing" the relationships between so many changing elements; it is only able to deal with subsets or chunks of a complex system.

Theoretical work and computational studies promised the power to see any number of relationships, to handle any number of variables; they would provide a way of analyzing the responses of a model circuit to different simulated perturbations. Those observations might or might not be confirmed in real, biological observations. Marder certainly didn't envisage computer or theoretical models replacing experimental work. She expected to use the two approaches in tandem and iteratively. Experimental data would inform the model, and the model would compute the behavior of a neuron in changing conditions. Its results, she hoped, would suggest new mechanisms that could be tested at the bench, and those experimental results might be used to refine the model and to explore the consequences of the mechanisms.

Marder was, of course, aware of John Hopfield's modeling of neural networks in the early 1980s. She had watched from the sidelines as colleagues such as Pete Getting, Brian Mulloney, and Allen Selverston attempted to find general rules about central pattern generation by developing computer models. Compared to today's computers, the power, speed, and memory available were pitiful. But, like her peers, Marder knew it was the only route to the core of the matter. She took the plunge in no half-hearted fashion. A period of extremely productive work, from the late 1980s to the mid-1990s, gave her lab its place in the discipline of computational neuroscience. It introduced a new hybrid of electrophysiological and computer technique now called the Dynamic Clamp and eventually yielded a surprising and beautifully formulated theory of neuronal stability. Along the way, in 1990, Marder became a full professor; she was forty-two years old.

* * *

Marder's starting point was an unsolved problem that needled her mind: the strange behavior of the anterior burster (AB), the pacemaker neuron. Her work with Scott Hooper showed that the neuromodulator proctolin spurred an isolated AB neuron to burst at two cycles per second but roused a sluggish pacemaker group as a whole to only one cycle per second, whereas the AB neuron would have been expected to set the pace. They had argued that the ventricular and pyloric dilator neurons, which were not proctolin-sensitive, were holding down the frequency of the AB neuron because of the electrical coupling between them (see figure 6.1). It was a dynamical argument that made intuitive sense but left Marder unsatisfied. It led her to think about neurons with oscillatory properties. How did they shape the output patterns of the circuit, and how were they in turn affected by it? In this case, how were the electrically coupled neurons of the pacemaker group acting together to control the frequency of the circuit?

The next step, she thought, was to investigate those oscillatory properties. She knew she had to look for people who had the expertise that she and her team lacked. Marder has a tremendous talent for exciting people intellectually so that they want to work with her; it's a special trait that attracts interesting people. She also has accurate intuitions about people, noticing a promising intelligence or well-focused curiosity where others might miss them. Because she has a firm sense of the direction her science should take, she has no fear of bringing new techniques into her work if they promise to further her understanding. Most heads of labs want to feel they're on top of the methods their teams are using, but Marder has the confidence that she'll learn enough to distinguish what she wants within the data. Now she started to build working relationships outside her lab and outside biology.

Marder had given a lot of thought to the kind of collaborations she wanted. She says she felt that what the theorists promised but usually didn't deliver was to reveal fundamental general principles. The general principles they were talking about were at a level that wasn't satisfying to Marder—they were too general. There was a yawning chasm between the terms in which the theorists expressed their principles and the detailed biology familiar to her. She was looking for some middle ground.

One of the first specialists in describing oscillators she met was the mathematician Nancy Kopell of Boston University. Marder started to pick her brains. "Obviously I wasn't trained in that kind of theory, and I was probably fairly incoherent about it." Kopell was working on the mathematical description of the lamprey swimming central pattern generator as a chain of coupled nonlinear oscillators. The action of oscillators in a closed system is relevant to the analysis of chemical concentrations as well as to neuronal circuits and in Kopell's lab, on sabbatical from Brandeis, was Irving Epstein, chemistry professor and leading light in the field of mathematical models of oscillating chemical reactions. These reactions display phenomena that can, with a few adjustments, be mapped onto biological systems, and Kopell suggested that Epstein might like to work with Marder on a model of a neuronal oscillator. When his sabbatical ended and he went back to Brandeis, the collaboration continued. Marder was feeling her way into computation and now thinks she didn't make the best of her early questions and conversations with Kopell and Epstein. She found she was learning quite a lot about building models but not getting answers to her most pressing questions.

* * *

In 1988, after years of debate and dispute, the U.S. Congress appropriated a budget for a 54-mile high-energy Superconducting Super Collider, a particle accelerator to be built in Waxahachie, Texas. The announcement came too late for Larry Abbott, a professor of particle physics at Brandeis. He had already realized that the next great discoveries in his field were likely to be twenty years in the future; the long wait for the means to test his best work would eat up the middle of his career. He knew he had to do something else. He turned his mind to the analysis of systems outside physics and chose "the neuronal network."

When I met him, he qualified that, saying, "I was just modeling neuronal networks—as physicists do."

I wanted to know what he meant: "You were building the early sort of neural model, where a neuron was either on or off?"

"Exactly, yes, and I wouldn't have called them really neurons. It was just a way of looking at systems, at networks."

"Neurons being thought of as conveniently binary," I suggested.

He grinned ruefully. "By physicists, yeah. Not by Eve."

In 1986, at a meeting for Brandeis faculty members to present their work to one another, Abbott talked about his abstract neuronal networks. Afterward, Marder went up to him with the inimitable invitation: "Come and see *my* neuronal networks."

Abbott didn't take up the invitation immediately. With his graduate student, Tom Kepler, also a physicist, he had started looking at memory storage in neuronal networks, and they were building electronic models, real physical ones, of small networks. They were also reading about neuronal network theory, which was mostly physics; early models such as the celebrated Hopfield model were indistinguishable from physics. Neither Abbott nor Kepler knew any biology, let alone any neuroscience.

One day, Marder's graduate student Jim Weimann took Kepler into the lab, and Kepler, intrigued, went to fetch Abbott. Mike Nusbaum showed him around. Nusbaum was doing an experiment at the rig, and Abbott saw that, instead of turning a knob on some resistors, you could turn a knob on a stimulator and see the real neuron in the ganglion reacting. He was enchanted.

Abbott says he instantly decided to switch disciplines. He went off to read up on some neuroscience. "After a while I went to see Eve, saying, 'I've got an idea—do you think it could work this way?' And Eve said the perfect thing: 'Well, we'll do an experiment this afternoon. We'll see if you're right.' And the next day she said, 'No, you weren't right.' And that had a huge impact on me, this back and forth over a day, that you could learn something just like that. I mean, in particle physics that process would take twenty years at least."

Over the next year, Marder and Abbott talked every day, usually for an hour or so, and Marder brought Abbott into neuroscience. He read the major texts and arrived every day with more questions. He learned dissections and did a bit of work at the rig. Marder talked biology to him. Abbott talked theory to her. They had both realized there was something important to be done, but Abbott thinks it was about a year before "We each said something and we looked at each other and said, 'What you just said makes sense to me,' and suddenly—Bam! We were communicating."

I asked Abbott whether he had been aware that Marder already had an idea at the back of her mind, whether he knew what she was looking for. He said, "I became aware of it pretty soon. Eve was one of the first systems neuroscientists to realize what theory was for. She knew exactly what it could do for her system way, way back before I ever met her. She knew what she was looking for scientifically. I just don't know how she knew that I was going to be worth it. Somehow she had the faith that it was going to pay off one day, and I don't know how she had that faith."

And so a famous collaboration began in a most unusual way. It would be hard to find another instance of a professor moving into a completely new field under the tutelage of a fellow professor. "Honestly, I wouldn't have done it if the tables had been reversed. If somebody came into my office, another faculty member, who knew as little physics as I knew neuroscience, I would have told them to get lost, and I think most people would too." But Marder had recognized in Abbott an exceptional intellect, a true respect for biology, and, of course, the skills and mathematical imagination that she herself would never command. The effort that Abbott expended in taking up the neuroscience challenge fully deserved her confidence, and their adventure together has been richly creative. Moreover, with hindsight, he had made the right call. In 1993, the Waxahachie supercollider was canceled by the House of Congress. Abbott is now the William Bloor Professor of Theoretical Neuroscience at Columbia University and one of the world leaders in computational neuroscience.

* * *

Marder could now work with Abbott, Epstein, and Kopell, accomplished colleagues who were interested in some of the problems uppermost in her mind. In 1989, the foursome wrote a collaborative grant application for dynamical systems study of small networks such as those in the stomatogastric ganglion. Talking about it, she is still wryly amused at their lineup of a biologist, a physicist, a chemist, and a mathematician, long before the word "multidisciplinary" became the cliché it is today. They got the funds.

Together with their students and postdocs, it was quite a complicated grouping, and over the next five years, they worked and published in various combinations. At first, with Epstein, Marder worked on oscillations from a theoretical point of view, describing an oscillator that resembled the anterior burster, although it was not based on experimental data but on the generalized characteristics of that sort of neuron. It highlighted the unexpected extent of the influence of calcium ion currents flowing into and out of the neuron, showing that even a slight change in these currents had a detectable effect. Kopell and Abbott worked, separately, on the more generic properties of oscillators. Kopell was still looking at neuroscience from the outside, as a source of fascinating mathematical problems. Kopell and Abbott then wrote a paper together. Abbott's student Tom Kepler finished his thesis on neuronal networks and joined Marder's lab as a postdoc in 1989. The group broke up when Epstein became Dean of Arts and Sciences in 1992. Marder inherited Epstein's postdoc Frances Skinner, who was interested in dynamical systems and authored a paper with Marder and Kopell. Abbott and Marder started working together more seriously.

Tom Kepler decided to build a model of the pyloric rhythm, and that suited Marder because she wanted a model of its pacemaker kernel—those electrically coupled oscillators, the anterior burster, and the two pyloric dilators. Marder wanted to make sure they had a convincing model for this curious and hard to explain trio before going on to the whole circuit. She wanted to see how the braking effect would show up in a model. She asked Kepler to build a model oscillator and then constrain it as though it had electrical coupling with a nonoscillator, to check that the nonoscillating cell slowed down the oscillator. Kepler did the first run of simulations and went to tell Marder that the nonoscillator sped up the pairing rather than slowing it down. Marder said he must have made a mistake. Kepler was offended and went off for the weekend. When he came back, he said there was no mistake, and he knew what was going on. He had found that there were two cases depending on the characteristics he gave the oscillator and the strength of the coupling.

Could there be two cases in the real world, of which the AB-PD cells demonstrated only one? "I was completely, totally, stunned!" Marder's eyebrows shoot up at the memory. "Because I didn't know there were two cases. I only knew about the oscillators that went faster when you depolarized them, and I didn't know there were others that could go the other way. I realized that the actual details of how that oscillator was built completely, qualitatively, changed its behavior in the network. It wasn't subtle, you know, it was all or none. It was either this or that. And for a biologist to deeply understand that you can be very close to a bifurcation between qualitatively different behaviors was astounding. Scott and I had implicitly generalized from the preparation we had in front of us to thinking it was always going to be that way. We were wrong, and that, the danger of a rushed generalization, was one really important lesson for me. The other was that laboratory experiments on preparations may only reveal part of the truth."

It was a strong proof of the value of computational modeling, and Marder was excited and encouraged by it. They had, as yet, no experimental data showing the second case, and Marder doubts that other biologists or electrophysiologists knew about it. It turned out that the stomatogastric ganglion contains an example of the other case, too. The dorsal gastric neuron (DG) is that kind of oscillator; many years later, when Marder saw the DG's behavior, she recognized it immediately. In fact, both conditions are fairly common. "But people often don't realize what they're looking at because, to see it, you have to actually look at the effect of current on an oscillator. And people don't always do that." Her earlier work with Irving Epstein showed that slight changes in calcium ion currents could affect an oscillator. With hindsight, she says, "He had already understood, better than I had at the time, that you could get bursting by two qualitatively different mechanisms. In the model that Tom was working on, it was just the balance between the calcium and potassium currents. Very simple. Scott and I hadn't realized it was just the current injection doing it; we thought it was somehow the current coming through the coupling. And the minute you saw it, it was just completely obvious what was going on. But nobody I knew was aware of it."

If the oscillator cell reacts to injected current by keeping calcium and potassium currents matching each other, then the oscillator can go faster and faster—until it crashes. But if the calcium current increases faster than the potassium current can keep pace, the oscillator goes faster and then slows down because the calcium current determines the duration of each of the neuron's bursts, and increasing the duration eventually slows the frequency.

Kepler, Marder, and Abbott published "The Effect of Electrical Coupling on the Frequency of Model Neuronal Oscillators" in *Science* in 1990. In the same year, Marder and Epstein's paper on neural oscillators came out. In the Marder lab's list of publications, 1990 stands out as a watershed year;

from then on, computational work takes its permanent place beside the experimental work.

At last, a computer model had opened Marder's eyes to a phenomenon she would not have suspected. She loved the paper "because I learned something from it. Larry dismissed it because the math was unexciting." Abbott thought it was trivial; he would always have assumed that changing the parameters on a model would change its behavior. What was this fuss about a couple of differential equations? But he realized it must be interesting because he saw that Marder was absolutely convinced of its importance.

Picking up the theme of that paper, Marder and Abbott, with Scott Hooper, wrote, "Oscillating Networks: Control of Burst Duration by Electrically Coupled Neurons," which came out in 1991. This time Abbott was pleased with the work because it provided a theoretical answer to a subtle biological problem: how phase constancy arises from currents with fixed time constants. At the time, Marder says, Abbott was particularly interested in the problem of phase constancy. "He was never interested in building detailed semi-realistic models. It just wasn't his way. He was really still a particle physicist." But Abbott had published his last paper on particle physics, also in 1990, and considers that that marked the end of his career as a physicist.

* * *

Remembering Marder's tale of woe about her mathematical abilities, I asked theorists who have worked with her whether she was limited in any way. If I hadn't told them, they never would have known. Frances Skinner replied, "Eve felt her limit in math? I didn't know that. I'm surprised because, from the very beginning, she went right to the heart of what the model was trying to do, how the model could be helpful, and forced me to explain it in a way she could understand. And I knew she was a biologist and not a mathematician so I remember being really impressed by that. I always wanted to combine biology with my applied math background. But in applied math, you're taught to non-dimensionalize your set of equations to try to reduce the complexity. So you work with a nondimensioned system, you don't have any units in your parameters. The first time I showed her something, she said, 'This doesn't mean anything, what is this?' She said we were not even talking the same language. You have to think what a model means with respect to the biological system. I tell that story to my own students."

"Eve helped me think clearly about that interface between math and biology. I got to do experiments in Eve's lab and learned a lot about the combination aspect, hands on. My own thinking has certainly been molded by her."

I asked Marder if, when she's looking at a model, she can judge whether it reflects a biological reality or whether it's wandering off hand in hand with the theory. She said, "I always know. I think I have a principled way of asking those questions in my head, but I think I do it pretty automatically. Obviously the gifted theorist is somebody who picks an appropriate tool for the question in hand. And that's one of Larry's gifts."

Then I asked her how she judged or interpreted models, thinking that she had to be able to read them at least. The fruit of hard-won experience and the assurance she has gained from it was evident in her answer, which is an important general lesson in itself: "I can't read code. I used to try and read the math a lot, but I don't tend to much now. I learned, over a period of time, to pay attention to what question people were trying to answer— if there was one—and what assumptions they were starting with, which sometimes they weren't aware of. And then to figure out approximately what they had done, and then whether that told me something at the end that I didn't know before. And over the years, I stopped worrying about all the stuff in between."

Laughing, Abbott told me, "You can't bluff Eve with the math any more. She just says, 'I know that doesn't make sense.' And she's right. Especially in collaborative science, it's got to make sense. You can't say, 'Oh this is too technical; I can't explain it to you.' You can say that about pieces of it, you can say getting from this equation to that equation was very technical. But if you can't explain it, you're in trouble." He says Marder has the confidence to just say, "I don't care how you got from that equation to that equation, you've got to tell me why you bothered."

* * *

Looking for the middle ground between the biologist experimenting with real neurons and the physicist soldering electronic components together, Marder and Abbott hit on the idea of "hooking up" real neurons to electronic circuits as a way of studying the dynamics of neuronal circuits. They imagined building artificial circuits from cultured neurons coupled up to electronics, and Marder dreamed up an artificial chemical circuit with chemical synapses, using the membrane potential of one cell to control an iontophoretic electrode that would deliver a neurotransmitter into a follower cell. She just couldn't work out how to build an artificial electrical synapse. At lunch one day, Abbott shrugged off the difficulty saying it was trivial, wrote out the circuit diagram, and they went to work. Marder gave

Abbott's blueprint to her laboratory technician, Michael O'Neil. "He was in my lab for maybe eight or ten years. He was a very smart guy, and he was a self-taught programmer. He left to start a company selling his software for managing research grants, and for quite a long time the biology and biochemistry departments here were running his software."

O'Neil designed and made the electronic circuitry that simulated electrical coupling between two real, biological, stomatogastric ganglion neurons. The analog circuitry compared the membrane potentials of the two cells and injected current (as a solution of ions) proportionally to the voltage difference between them. The strength of the coupling could be varied by the experimenter, in this case, a graduate student, Andrew Sharp. The circuitry could also be set to trigger Marder's artificial chemical synapse onto the "postsynaptic cell" at a pre-determined point in the oscillatory phase of the "presynaptic" one. The individual cells were in separate dishes and could therefore be subjected to different pharmacological environments. The study showed the related roles of membrane potentials, intrinsic properties of neurons, and the strength of the electrical coupling between them. It confirmed the earlier results of both Hooper and Kepler, showing the two effects of a passive neuron on an oscillator electrically coupled to it: speeding it up or slowing it down.

They submitted the work to *The Journal of Neurophysiology* early in 1992, thinking it was a first. But they were quickly made aware that in 1990 and 1991, in the far distant field of cardiology, Joyner, Sugiwara, and Tau had published reports on isolated rabbit heart cells coupled by a variable resistance; they had called their analog circuitry a "coupling clamp." An acknowledgment had to be added in proof, and this minor incident is without consequence except that it seems to indicate the parentage of the term "dynamic clamp."

The dynamic clamp now has a place in the armory of electrophysiologists all over the world. Marder's lab developed and refined this technique, and it has burnished her reputation and that of her lab. The concept matured quickly and benefited from a short conversation with a mathematician. Marder had given a talk about the artificial circuit work at a Computational Neuroscience course at Woods Hole. Afterward, John Rinzel remarked that the model's electrical synapse could be programmed to put any conductance you wanted in the follower cell. "Yes, indeed!" thought Marder, and when she told Abbott, he simply said, "Of course." And so they set out to replace the analog electronic circuit with a computer program. Again, Abbott designed it, and O'Neil wrote the code and made the equipment. This time it was a hard and lengthy process because they were at the cutting edge of what computers could do. The principal problem was speed. The program had to be written in machine language so that it could run fast enough to work in real time, the biological timing of the real neurons. It was only just feasible. Andy Sharp test drove each step and found the snags before settling into experimental mode. It was months and months of work, and it created unexpected problems in Marder's lab.

Sharp, naturally a rather intense young man, was working on the dynamic clamp with O'Neil, whose program was written on a PC. A few yards away, another team was aiming for the same goal, writing code on a Mac. Gwendal LeMasson, a postdoc from France, a physician and self-taught computational neuroscientist, was trying to do something even more ambitious than the PC group. His teammate—and wife—Sylvie LeMasson was a computer scientist. Their idea was to build a multiuse simulator that would run the dynamic clamp and also be a modeling program. The two groups refused to work together, and Marder says, "It was not nice."

A feature of Marder's lab, mentioned by many of the people I have talked to, is that there is no unhealthy jealousy between groups. Certainly, there have sometimes been strong personal clashes—scientists aren't angels—but nothing like the outright bench wars that break out in some labs. Occasionally, Marder has had to move people to different offices and reallocate space. Once or twice in her lab's history, she told me, "I came in in the morning just hoping they weren't going to kill each other."

This time there was also a simmering dispute at the rigs, where Sharp was sharing uneasily with a postdoc whose experiments required a different configuration from his. Sharp taped down a frontier. Gina Turrigiano ignored it as being impossible in practice. At that point, Marder moved Turrigiano to a rig of her own. Across the lab, Turrigiano was sharing office space with LeMasson and getting on with him very well indeed.

In the end, in 1993, the work Marder published first was with Sharp, O'Neil, and Abbott. It was a short introductory description of the concept: "The Dynamic Clamp: Computer Generated Conductances in Real Neurons." This paper was followed by a full presentation with almost the same title: "Dynamic Clamp: Artificial Conductances in Biological Neurons." At the end, the acknowledgments included, "We thank J. Rinzel for early discussions of this idea, and R. Calabrese and M. Nusbaum for helping us troubleshoot the system. S. Renaud-LeMasson and G. LeMasson collaborated with us on the parallel development of a similar system simulating full model neurons."

In the same year, *Science* brought out an important paper by LeMasson, Marder, and Abbott that set off a train of discoveries and ideas. With everyone's work published, peace returned to the lab.

The dynamic clamp was the marriage of experimental physiology with computer modeling. It was already standard practice to inject a current with a controlled waveform and time course directly into a real neuron. But now the dynamic clamp could alter that injected current according to real-time feedback of the cell's membrane potential—its state of excitability—in an iterative process.

In its simplest form, using a reference electrode and two intracellular electrodes, one intracellular electrode reads the membrane potential in real time to the computer, and the other injects the current (the number of charges per unit time) computed for the particular experiment, establishing a loop between the two. The real neuron, of course, has other types of ion conducting channels in its membrane, and these interact with each other and with the "artificial channel" and are affected by the resulting changes in membrane potential, the dynamics of its values (see figure 7.1).

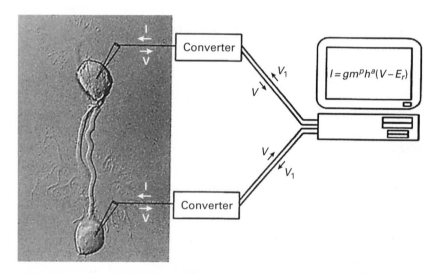

Figure 7.1
Micrograph and diagram showing the principle of the dynamic clamp. An electrolyte-filled microelectrode penetrates each of two cultured neurons in a Petri dish. The electrodes measure membrane potential (V), and this is converted from analog to digital before reaching the computer. According to the desired manipulation, a conductance is computed and converted to an analog voltage controlling the injection of current (I) into one or both neurons. (From Sharp et al., 1993.)

The dynamic clamp can also be used in hybrid networks, where a biological neuron and computer-modeled neurons interact in real time. The mathematically defined model neurons can be manipulated, and the interplay with the responses of the real neuron can then be monitored. As a tool, the dynamic clamp opened up rich possibilities for electrophysiology. Mike Nusbaum describes it as, "really the physiologists' equivalent of a molecular geneticist doing knock-outs and knock-ins. It came out of the need to try and manipulate the neurobiology in a certain way. When she first described the idea, it seemed so obvious, like, why doesn't this already exist? And yet of course it had not been thought of, or anyway not implemented."

Marder now says, "It was actually a method that was destined to be developed. A look-up table version of a sodium current was developed at about the same time by Hugh Robinson at the University of Cambridge. That was initially faster than our earliest version but not programmable in the same way."

Nusbaum adds, "Eve was excited and already had a list of questions— quite long—well before they had actually developed the first dynamic clamp apparatus enough to try some experiments."

Nevertheless, the dynamic clamp started with two disadvantages: it was slow compared with the speed of a real biological process, and the effect of its current injection was limited locally to a small area around the electrode. Computers in those early days were slow, so only relatively slow currents could be implemented. But researchers in Marder's lab, then in others, soon started tweaking and patching the codes, while computers continued their rapid increase in speed. The dynamic clamp improved and with it the array of simulated perturbations that could be studied in a biological timeframe. But the limitation due to the restricted localization of the calculated conductance near to the recording electrode remains.

* * *

As Marder was embarking on the early computation work, the first attempt at building a model based on real experimental data was underway at the other side of her lab. Marder approached this project serenely, with no idea of the difficulties it would create for her. It started a long story that resonated through her work for years, during which at least one of her suppositions about the neuron changed dramatically, indeed to an opposite view.

In 1986, Marder had taken on a graduate student from Chile, a biophysicist. Jorge Golowasch was another ion channel enthusiast and wanted to identify and measure the flow of different ions they allowed to pass through the cell membrane. Marder sent him to Cold Spring Harbor for a course on electrophysiology, particularly the technique called "voltage clamp," which is used to understand ion channels. It prevents unwanted change in the membrane potential of an area of membrane under investigation. The membrane potential, which is usually expressed as voltage, is controlled by the researcher who can "clamp" it at any desired value and can thus measure ionic currents as a function of voltage.

Golowasch chose to work on the lateral pyloric neuron in the crab, *Cancer borealis*. There is only one LP in each stomatogastric ganglion, and it is among the easier of its neurons to identify. It is a motor neuron, exciting muscles that constrict the pylorus. It fires in alternation with the two pyloric dilator neurons, its constricting muscle action in antagonism to their dilating action. The LP is a typical motor neuron of the stomatogastric ganglion in that it also makes inhibitory connections in the neuropil, so it has a role in sustaining the pyloric rhythm. It was known that several neuromodulatory substances influenced the LP, altering the number of action potentials it produced per cycle of the pyloric rhythm. This variable excitability arises from changes in the neuron's membrane potential.

An explanatory aside may be helpful here. A neuron's membrane separates the number of positively and negatively charged ions on either side. A neuron that is not in an active or perturbed state has negative charge inside relative to the extracellular milieu. This separation of charges creates a difference in electrical potential across the membrane; the potential is changed when membrane ion channels open, allowing ions to pass into or out of the cell, or close, preventing their passage. The flow of ions, say potassium ions (K^+), across a cell membrane constitutes a current, in this case, a potassium current, that depends on the opening and closing of channels specific to those ions. A conductance in a neuron is the measure of how easily such an electrical charge moves across the cell's membrane— the converse of electrical resistance—and is also specific to each ion. Think of these channels as "ion conduction channels." Then "maximum conductance" for an ion is reached when all of a cell's ion conduction channels specific to that ion are open, allowing the maximum current of those ions to pass through.

Golowasch wanted to characterize all of the currents affecting the LP neuron. The project interested Marder, and not just as an exercise collecting information for its own sake. Like other neuroscience researchers, she wanted to find out how a neuron's conductances influenced its membrane properties, its excitability, and therefore its activity patterns, under different neuromodulation conditions. The prevailing paradigm was that somehow the conductances would be the key to understanding a neuron's activity in the circuit. As it turned out, Golowasch was able to identify and describe six currents, three outward and three inward. Most interestingly, the ratio of maximum conductances of the three outward currents measured in any one cell was similar to the ratio between the average values of these conductances from many different LP cells, implying that a balance to which all LP cells are predisposed necessarily exists between the conductances.

The relationship between these conductances and the resulting activity of the cell could not be described as linear. Marder saw that the problem lent itself to computer modeling; they would build a model of the dynamics of the LP neuron. She wasn't aiming to use the model to test a theory—she had none. She thought a model would tell her more about which currents were important for which state of excitability of the LP, while taking into account all its active currents at the same time. It was impossible to do that experimentally; only computers could handle the many variables. The next question, and a thorny one at that, was what to feed into the computer.

Golowasch worked on the model with Frank Buchholtz, a postdoc in Irving Epstein's lab. It wasn't easy. Remember, we are talking about the computers of the late 1980s. The model was written in Fortran, and it ran on the university's mainframe computer. Making changes to the model was cumbersome in a way that most of us have now gratefully forgotten; each alteration meant taking the program back to the computing center for another run and waiting until the next day for the results.

The model was, of course, simplified in many ways. For instance, it pretended that the whole LP neuron acted as a single volume, and it did not take into account the length of the axon or variations in its diameter. Differential equations described the voltage, time course, and calcium ion dependence of each of the six currents. The model started from the experimental measurements that Golowasch had made, and the six currents were related to each other by a single equation that I throw in here just to respect the grand publishing tradition of permitting one equation per book:

$$c_m \cdot dV / dt = i_{ext} - \sum_j ij$$

Here, c_m is the membrane capacitance, V is the membrane potential, t is time, i_{ext} is the current applied experimentally, and the sum is that of the ionic conductances, which were each described by a suite of equations founded in the data. I won't do it again, but that's the sort of thing that makes up computer models of neurons.

It should then have been possible to simulate electrophysiological experiments, perturbing the model neuron as one might the neuron itself. But that didn't happen immediately. Two major obstacles haunted Marder for years, even after the model had been patched up and made to work.

First, they found themselves endlessly tweaking and re-tweaking the parameters because the model played up and required experimental data they found they couldn't measure in the LP preparation. It was a soul-destroying process. They spent months "hand-tuning," changing one of the currents, of potassium say, which meant the calcium current had to be changed—and then the sodium current would be wrong. They couldn't measure the sodium current at the rig, and the model turned out to be dependent on it.

The second problem was the choice of data. If Golowasch had, for example, measured the A current (an outward potassium ion flow) in fifteen cells, then they had to decide whether to use the mean of the fifteen values or the "best" value. Best, when you're talking about currents to a biophysicist, means fastest and biggest because everyone knows a poor voltage clamp gives something small and slow. Marder told me, "The question sat there as a piece of extraordinary discomfort in my head for many years. I knew I didn't have a rational way of answering that question. I think we ended up using a combination, means for some things and best for others." This was a subjective assessment in my opinion, and she replied, "Well, yeah, it was completely arbitrary. And I knew it was, and it drove me crazy."

Eventually—after nearly a year's work—they had a model that Marder could describe as "sort of okay." In fact, at the time, it was even one of the better semi-realistic conductance models around, although they got it to mimic the behavior of a real LP neuron only by manipulating values for the membrane capacitance and the characteristics of the sodium currents. They learned from this tuning that the cell's behavior depended most sensitively on the set of maximal conductances—the model would crash at the slightest error in that set. Obviously, it raised the question of how real neurons could establish and keep the right set of conductances to maintain their activity. "I kept on saying, this is too hard. Cells can't be doing it this way. Building this model, we were hostage to things we couldn't measure at all. And then with things we *could* measure, we didn't even know whether to use the mean or the best. And from that we were going to try and extract something about how the real cell behaved?"

Marder started talking to other people who were trying to do similar things. All their models were fragile; they crashed all the time, and everyone was fine hand-tuning. But most of the people building models were theorists who were not themselves collecting the data. It was an important distinction. Marder says theorists were less bothered with the vagaries of the data than she and her team were. "At the time, people who wanted to build a model of a hippocampal cell might find they needed a calcium current, so they'd go to the literature and find a calcium current in the cerebellum in a different species at a different temperature. Or they needed an example of a sodium current, so they'd find data from the crustacean walking leg axon and put it into a model of a mouse cortical neuron. I'm not making this up!" These compromises were necessary when the appropriate data weren't available; it was a question of making do with what was in the literature. But Marder saw that such fudges might well render a model useless for biology. Although she would often be frustrated in getting exactly the right data, Marder was aware of the importance of keeping the distance between model and biology as short as possible. From her long experience of observing one nervous system in one order of animal, she knew that the differences between species could not be so lightly ignored. "By the standard of the time, we were doing pretty well. At least we stuck to data from crustaceans."

Golowasch left Brandeis in 1990 with a fine thesis and his doctorate but without finishing the papers for publication. He went to a postdoc position at the Laboratoire de Neurobiologie in Paris. After a while, he sent his manuscripts to Marder. She knew he was in a fix because she herself hadn't been able to work out what progress they had made in the whole grueling exercise. So she got on a plane and spent a week in Paris.

"She went to buy lots of shoes. Her alibi, she said. We spent hours and hours working on the manuscripts. She would drive me mad." Golowasch shakes his head at the memory. "She would completely destroy what I had written and rewrite it, and I always thought, jeez, why? It wasn't so bad. I learned how to write papers thanks to that week."

But Marder, for once, had been stumped. "I found writing the discussion sections of those papers absolutely excruciating because I couldn't see what we had learned from doing this. I mean, I worked really, really hard to try and figure out what we had learned. We had learned some things, but they were little things; they weren't big things, and I kept on saying, 'How is it that we learned so little after doing this whole thing?'"

The work was presented in three papers published together in *The Journal of Neurophysiology* in 1992. The first described Golowasch's experimental results. Commenting on the similarity of the maximum conductance ratio for the three outward currents in a particular cell to that of the average of maximum conductances from the many different cells investigated,

Golowasch and Marder wrote, "This indicates that these currents adjust themselves to a constant relationship characteristic of the crab LP cell so that the LP's identity is reflected in the specific balance of conductances that produce its unique patterns of activity." That sentence would not have surprised anybody at the time because it was accepted that each type of neuron had a characteristic, describable set of properties specific to its type. I quote it because, in the rich turbulence of ideas that was just around the corner, this paradigm would be upset.

Today, Marder says, "Obviously the glimmer was there. So it might have already been there in my head or Jorge's. It was me musing on something I had noticed. Interesting—there's nothing like going through the records. What that really should have caused us to do, and didn't, we actually should have plotted the correlations." Seven years later, prompted by the results of one of the lab's later, improved computer models, they came back to these results and did just that. It was a revelation, and by then Marder was ready to integrate those results into the major upheaval of received wisdom she had initiated.

The second article described the structure of the model, noting the difficulties in building it but pointing out that, "The model LP neuron behaves in a manner gratifyingly similar to that of the real LP neuron." The third paper reported the investigations using the model and compared the effects of manipulating individual currents in the model neuron with the known effects of pharmacological interventions in biological experiments.

The articles clearly stated the objective to which the project was intended to contribute: "We eventually wish to explain how each of the membrane currents in a neuron that is part of a functional network contributes to its dynamical properties. In other words, we wish to understand, for each neuron in a network, how its activity within the network arises as a function of its voltage- and time-dependent currents and its synaptic and modulatory inputs." But asking how activity was shaped by membrane currents turned out to be the wrong question. Trying to interpret their results in those terms was probably the reason any useful conclusion escaped them, and Marder felt that no wider lessons had been learned.

Nevertheless, the intense thought that went into the effort to make sense of the exercise was to bear extraordinary fruit. The final discussion ended with a speculative paragraph that shows Marder feeling her way to new and deeper questions. It commented on what would eventually become a far-reaching examination of animal-to-animal variability and proposed two explanations. The first was the generally accepted one of experimental error: "The process of model construction forcefully reminds the physiologist of many unanswered problems. Specifically, there is intrinsic variability in all biophysical measurements ... However, biological neurons retain with extreme fidelity their characteristic electrical signatures. This may mean that most of the variability in physiological measurements is induced by the experimental procedure." The second presaged the ensuing direction of her work: "Alternatively, neurons may have interesting compensatory feedback mechanisms that allow them to produce their essential patterns of activity despite modifications in some aspects of their currents. If so, this indicates the existence of important cellular processes not taken into account in this or similar models, as models are not often extremely robust with respect to parameter modifications."

Having earlier concentrated on how neuromodulation changed the behavior of neurons, sometimes quite dramatically, now she was confronting the fact that neurons somehow achieved enough stability to maintain useful behaviors. She turned the conundrum over and over: "How could it take nine months for a smart postdoc and a smart graduate student to know how to build this model, and the LP cell gets it right every day? I knew we were missing something very important. And it was the fact that we hadn't learned very much that led me to call into question whether we were defining the problem the right way."

Marder says the papers written with Golowasch were, in themselves, not substantial scientific steps forward, but their intellectual impact on her was another matter. By the time they had finished editing them, they had framed the question posed in the last paragraph: "Our work thus highlights a fundamental puzzle of neuroscience: if the electrical activity of a neuron that confers on it its essential characteristics is so tied to the correct balance of many different conductances, how are these regulated so that the neuron retains its physiological identity throughout its lifetime, despite protein turnover and growth?"

I asked her whether different possible explanations were playing out in her mind, and she caught me out with the reply: "This is what you don't understand about how my mind works. I didn't spend much time on any detailed thinking because there was nothing to think about. It was a puzzle. Some people build up really complex edifices, but I don't. I'm very data driven. I have a very, very, deep, abiding respect for actual data, whether it comes out of an animal or a computer model. So when I see things that confuse me, I know they confuse me and sometimes I say, 'Oh, it would be helpful to see this so maybe we could understand that.' But I don't build complex hypothetical stories about the way it might work."

Nevertheless, Marder did think that the LP cell could only get the right balance of conductances by having all of its ion channels tuned together; it couldn't be balancing them one by one. Everywhere else, people were studying the regulation of ion channels one type at a time. She couldn't see how to put it all together. And so she started to pester Larry Abbott to come up with a better way of describing the processes. She now knew exactly what was missing and what she didn't understand. She says that, for an experimentalist, the key to working with a really talented theorist is to be clear about what you don't understand and to frame a coherent question that the theorist can think about. "I said something very simple like, 'the activity depends on all these conductances and the cell isn't counting sodium and potassium channels so how is it getting all the channels correctly balanced?' Larry kept saying the problem was too hard."

Eventually, Marder says, he just got tired of hearing the same question again and again. "So he sat down and thought for twenty minutes. But it must have been incubating because I was nagging him. Then he saw the key to it and probably took about three hours to outline it. He told me the answer before he did the model. There was nothing hard about it. Once you say it's a negative feedback system, you just have to decide what rules you're going to use for those negative feedbacks. Now you could easily say that as a trained biologist, a trained physiologist, I should have been able to see that. I didn't."

Abbott's important—fundamentally important—insight was that the activity of the neuron should be the controlled variable, not the activity of any one kind of ion channel, nor of multiple kinds of ion channels. In short, it was the opposite of what neuroscientists, including Marder, had been expecting. The cell's activity would depend on the number of all of the ion channels working together, and therefore activity could be used as the controlled variable, but feedback from activity would regulate the cell's conductances.

Abbott's plan was to model a simple negative feedback system, homeostatic in the physicist's sense of that word. He had been working on a paper about the regulation of burst phase, and he had realized that the calcium slow wave could be thought of as carrying the history of a neuron's previous level of activity. So when he finally listened to Marder, he said, "Oh, it's just all the same thing. I'll use calcium." Calcium turned out to be the best proxy for activity they could think of, although they never thought it was necessarily calcium and only calcium. Unlike the intracellular concentrations of sodium and potassium ions, which are high relative to their currents through the membrane, the intracellular concentration of calcium ions is low, and it can therefore be taken as a sensitive reporter.

Abbott started work on the model with Gwendal LeMasson. It came together quickly. Their starting point was the model of the LP neuron that Golowasch and Buchholtz had labored so hard to produce. The new model took into account seven maximal conductances. If these, or the modeled ion concentrations surrounding the model neuron, were changed, then the neuron's behavior changed. Behavior in this context means that the equations resulted in membrane electrical values that corresponded to burst firing, to low- or high-frequency tonic firing, or to "silence." The concentration of calcium ions inside the model neuron was taken to represent levels of recent activity: as in a biological neuron, more activity led to a rise in the calcium ion concentration, and less activity reduced it. Reading the intracellular calcium ion concentration as the feedback sensor, the model neuron responded to perturbation by adjusting its conductances (representing the number of each type of open ion channel in the membrane of a biological neuron) in a manner tending to restore its initial activity pattern, which was a single equilibrium fixed point in this case. It was a speculative model that presented a mechanism, at least theoretically possible, by which a neuron might regulate its conductances—that is, maintain an appropriate complement of ion channels for its function.

By mid-1992, when Marder saw the results from the model, she recognized that the mechanism it suggested rang true to her. "It was in my head. It was all there then." She was indeed well primed to appreciate the meaning of the model. Ever since the exasperating work on the model of the LP cell with Golowasch and Buchholtz, she had been considering the implications of those unresolved difficulties. "Thinking about homeostasis comes out of that because what you learn when you try and hand tune a model is that if you've got five or six different currents, the answer is the *relationships* among them, not the values for each. My frustration with that exercise was incredibly important." The problem had churned in her mind and had prompted the search for the right question. The new model indicated the answer: at any one time, a neuron's set of conductances, and therefore its membrane excitability, is regulated by its history of activity. Marder always refers to it as the first homeostatic model, although the word "homeostasis" is nowhere used in the published paper describing the work. She traces the seminal reworking of the nineteenth century concept of physiological homeostasis, that staple of biology exam questions, to this model. "It was very clear. I mean, the paper trail is very clear. That 1993 paper suggested, had embedded in it, many specific predictions."

The report on the model was unshowily called "Activity-Dependent Regulation of Conductances in Model Neurons." Marder says, "The answer's so obvious now, but it was not obvious then because that wasn't the way people were thinking. You know, that paper, that 1993 paper, really is a paradigm shift."

I asked her whether it had been recognized as such.

"No. It was published in *Science*, but it did not become important for a long time. It gets cited more now than in the first few years after it was published."

In retrospect, the abstract paragraph, the introduction, alone is dynamite, and it's hard to see how the paper could have been ignored. It starts, innocently enough, by pointing out that the electrical activity of neurons has to continue throughout an animal's life, even though the membrane and its channels are constantly degraded and rebuilt and even though the neuron's extracellular environment varies. Then the message: a brief description of the model simply states that the conductances of ion currents, that is, the number of open ion channels specific to each ion, depend on the model neuron's internal concentration of calcium ions, which, in a biological neuron, is related to activity. Thus, the model neuron's activity regulated its membrane properties.

Further, and this surprised the Marder team, the model suggested that different activity patterns in neurons' recent history could lead to different electrical characteristics—neurons might differentiate from other neurons of the same type if the inputs they had recently received were different. If two model neurons, described mathematically in exactly the same way, were each put through a perturbation, although not the same one, then at the end both returned to their original firing pattern. If the same manipulation were performed after adding an electrical coupling between the two neurons, again each receiving a different perturbation, then the neurons started burst firing in synchrony but with different intensities. But when the electrical coupling was cut off, neither neuron resumed exactly its earlier firing pattern, nor were the two final patterns the same. The equations describing the neurons remained the same, but now two different behaviors were being produced. The finding was reported with the laconic comment that the feedback system "can at once stabilize neuronal function and serve as a differentiation mechanism." These ideas promised a radically new perspective in the study of neuronal dynamics.

Unusually for Marder, theory had run ahead of data, and she was happily surprised not to be shot down in flames when the paper was published; she had fully expected the experimental biology community to criticize the work as being unrelated to biological data and, of course, not backed up by experimental work. "I wasn't worried about the theory community at all; I was worried about whether the experimental community would view it as too much fantasy world to take it seriously. It was really a very speculative model because at the time the rigorous biophysicists were obsessively tracking their channels around, and biologists were focusing on the molecular biology, so there was a lot of missing stuff in there. Experimentalists have this literal side like, 'Oh, this model can't be exactly right because you don't know what the calcium binding protein is.' I worried they wouldn't understand the utility of a model that was way ahead of the data."

The biological data that corresponded to the two major clues brought up by the model would be found surprisingly quickly, and in Marder's lab. A vigilant reader might have noticed this phrase in the paper's penultimate paragraph: "… experiments inspired by our model that use stimulation of cultured neurons are underway," with a reference to unpublished results from the Marder lab, the first author's name being Gina Turrigiano.

8 Tuning to Target

Marder made her cry at their first encounter. Gina Turrigiano was beginning her third year of postgraduate research in Allen Selverston's lab in San Diego. She was presenting a poster at the Society for Neuroscience meeting in 1984. Research leading to the award of a PhD is supposed to be original, but Turrigiano had just found out that someone else had done the same work in another lab. It was a near disaster for her, and she knew she would have to scramble to find and complete a new research project. Marder came up to look at the poster, immediately recognized the work, and laid into her. Nowadays, Turrigiano can laugh about it: "I think Eve was mortified that she had done this, so after that she took an interest in me. Her ideas were very exciting to me. At that time, Eve was articulating the notion of multifunctional networks, and it was a distinctly new way of thinking about how circuits worked. People still thought of circuits as a static kind of entity. Her work was really changing that, showing that the same group of neurons could be reconfigured in different ways to generate different behavioral outputs. It seemed to me it had to be true and it had to be something fundamental about the way all brains would work, not just the lobster's stomatogastric ganglion. And despite the somewhat rocky beginning, she has been one of the most truly supportive people, both personally and scientifically."

Turrigiano joined the Marder lab in 1990. In her customary way, Marder left her new postdoc the freedom to choose a research question. "Eve was very hands off—she didn't say, 'Here's an idea, go try to test it.' Not at all. She would give you a very basic prompt—I think it was, 'Go see if you can get these neurons growing in culture and then you can do interesting things with them.' So I started pulling out neurons."

Working on the Pacific spiny lobster, as she had during her work at the lab in San Diego, Turrigiano identified neurons in the stomatogastric ganglion, took them out and cultured them, and then started to investigate their responses to different neurotransmitters. She found that, even when isolated and plated, the cells retained the characteristic responses of their cell type, indicating that they kept up the appropriate receptors in their membranes. It was a substantial proof of the stability of the so-called "determination decisions" in early development that lead to a cell acquiring its identity and its specific set of receptors. A pyloric dilator (PD) neuron, for example, would have certain kinds of receptors for certain substances and none for others, and apparently this decision was stable—once a PD, always a PD.

When Gwendal LeMasson arrived as a postdoc in 1991, he was given office space with Turrigiano. Naturally, they talked about their work, and Turrigiano was fascinated by the idea behind the model he was starting work on with Marder and Abbott: that a neuron's recent history of activity might somehow act on the ion channel populations in its membrane, thus influencing its electrical properties.

Turrigiano wondered whether she might see something like that in her isolated neurons. Following the neurons more closely than before, Turrigiano saw that, over three or four days, they became intrinsic bursters regardless of what type of cell they were or their firing pattern when undisturbed in the ganglion. There, most of them would be driven to maintain a characteristic firing rhythm as they were alternately inhibited and released from inhibition in the circuit. In the ganglion, some of them might burst in response to a neuromodulatory substance, but in culture there were no modulators to influence them. Surprised and intrigued, Turrigiano made extensive recordings and found that the balance of inward and outward ion currents had changed to reach this bursting state. "I was finding things in culture that we had not expected. I don't know if I would have seen the experimental results without that theoretical framework—would I have interpreted it in the same way? Who knows? These cells were becoming intrinsic bursters. I was really excited, seeing this phenomenon happen in culture and realizing that it was actually a manifestation of the idea that Gwen was working on."

Turrigiano supposed that the absence of rhythmic drive might be causing the neurons' intrinsic properties to change, inciting them, in effect, to look for their normal activity. Her prediction was that, if she restored the inhibitory drive the cells normally received in the ganglion, they would go back to their tonically firing state. So she stimulated the cells with long, inhibitory, rhythmic pulses, making them burst on the rebound from them. About an hour after she stopped the stimulation, their intrinsic properties had changed, and they had transitioned back to tonic firing. "You could just flip these guys back and forth in the course of a couple of hours. I think the first time I showed Eve, her jaw dropped."

This phenomenon was so unexpected—and almost too opportune—that the confirmatory experiments were much fussed over before the work was written up. As well as providing the experimental confirmation of the computational model, the latter's use of calcium ion concentration as the feedback medium was supported; the changing patterns of activity in the cultured neurons were dependent on the extracellular concentration of calcium ions, and activity could be prevented by calcium channel blockers. Marder quite reasonably thought the paper should be sent to *Science* as it followed up and confirmed the theoretical work the journal had recently published. But *Science* rejected it on grounds that nonplussed her: "They said it wasn't sufficiently novel because you already proved this in the model! I don't usually argue with editors or rejections, but I called the editor, asking, 'How can you conceivably reject an experimental paper because a model predicted the phenomenon? This is actually the proof of the model. This is the biology. The model's only a model.'" *Science* published the paper in 1994.

The paper suggests that stomatogastric neurons "adjust their conductances in a homeostatic manner," but otherwise the word "homeostasis" is neither mentioned nor discussed. But Marder's conviction that modeling would allow her to articulate new ideas—and her dogged pursuit of the right question—had revealed a phenomenon that had been missed in experimental work—namely, that neurons could return to a functional state after acute perturbation.

An obvious reason for missing it, obvious once you see it, is that many experimental interventions impose a rapid timescale for neuronal response that does not allow for the much slower pathways that compensate for perturbation. Slow compensation mechanisms manifest themselves over hours or days and would be overlooked in many protocols.

* * *

To Marder's satisfaction, the theoretical and experimental work now advanced in parallel. Marder and Abbott embarked on a new model with a gap year visitor, Micah Siegel, who was working in the lab after graduating from Yale and before starting his PhD at CalTech. This time the model took the neuron's shape into account; it was a "multicompartment" model, which considered the neuron as having a soma compartment with various neurite compartments projecting from it—at its simplest this can be thought of as the position of the compartment relative to the cell body, its size, and, in the case of neurites, their diameters. The morphology was based, not on a stomatogastric neuron, but on a much-studied type of neuron in mammalian brains, the CA1 pyramidal cell of the hippocampus, for which experimental results were available as comparators. When the model neurites were stimulated in compartments along their length, they developed different levels of conductance from those found on the model cell body or on neurites projecting from the other side of it. (See plate 7.)

Thus, the model indicated that the population of ion channels could be distributed unevenly, with locally different densities, depending on both the electrical activity of a compartment and the cell's morphology. Marder had long suspected that the electrical properties of a neuron, at any one time, were determined not only by the numbers of each type of ion channel present and their state of open or closed conductance, but also by their distribution in the membrane. This sort of clustering of voltage-gated ion channels would soon be demonstrated experimentally in other labs working on hippocampal neurons. Within ten years, it would be generally accepted that mRNA is transported from the nucleus to neurites and that localized protein synthesis is, at least in outline, one of the mechanisms leading to the local differences in channel abundance suggested by this model. Again, the word "homeostasis" does not appear in the published report, but the balance of these clusters could be interpreted as a stabilizing factor, perhaps evening out localized extremes of synaptic input.

The lab's next model was based on Gina Turrigiano's experimental data. Turrigiano and LeMasson built a model of the behavior of the isolated stomatogastric neurons in culture—a single-compartment model because the pulled neurons were effectively reduced to one compartment, the soma. It confirmed the experimentally observed increase in inward and decrease in outward currents, as well as the resulting behavior of the biological neurons (figure 8.1).

In the process of extracting a neuron from the ganglion, neurites were inevitably cut off, but after a couple of days in culture, some regrowth could be seen. The theoretical confirmation of the altered currents was therefore especially important because it showed that the increase in inward currents found by Turrigiano was unlikely to be the trivial result of the added membrane, with ion channels, created by this growth of new small processes.

Marder was pleased with the paper they wrote. "What Gina found in her 1994 *Science* paper and in this paper was that all cells, regardless of their cell type, revert to this sort of generic wanting-to-burst mode of activity

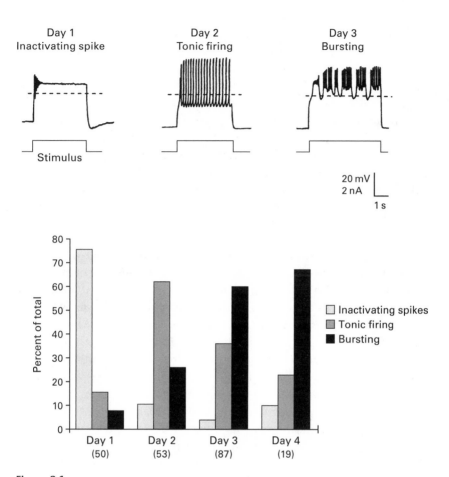

Figure 8.1

The upper panel shows the activity states of an isolated neuron responding to a stimulus over three days in culture. The dotted line is at −40 mV. The lower panel shows the percentage of neurons (total number in parentheses) in different activity states over four days isolated in culture. (After Turrigiano et al., 1995.)

after major perturbation. But her 1993 paper had shown that in culture, when isolated, identified cells retain the receptor specificity of their type. That makes some sort of sense: the generic excitability and the receptor specificity fit together." The discussion concluded that the loss of drive (the rhythmic inhibitory stimulations in the intact circuit) when a neuron was isolated contributed to the adjustments seen in its conductances. And as predicted in the earlier models, the balance of inward and outward currents in the model neurons was regulated by the level of activity.

Taken together, their investigations indicated that the excitability, the electrical state of a neuron's membrane, could change—not because the cells were expressing different ion channels, but because they were adjusting, removing, rebuilding, or synthesizing more of what they had always had in their membranes in response to their recent history.

From these several results, in both model neurons and real ones, Marder knew that the principle of homeostasis in the self-regulation of neurons was sound. "It was completely obvious, once you accepted that the pattern of activity to be maintained constituted the homeostatic set point, then there were consequences. And there were experimental things to do and there were theory things to do."

* * *

Many of the investigations that Marder wanted to carry out were impossible at the time. She wanted to build a new multicompartment model of a neuron more realistically extended in space to examine how it managed to localize its ion channel densities. She wanted a detailed model with DNA and RNA pathways and with realistic rate constants. So far, the treatment of the rate of activity-dependent modifications in the models was unnatural. In the animal, this regulation has to be slow, relative to the action potential firing capacity of the neuron; the adjustments occur on the scale of minutes, hours, or days because of the molecular pathways necessary in a biological neuron responding to perturbations. The lab's models used a much shorter time scale. But, such fantasy models would have been impossible to build then; nobody knew enough, neither biologically nor theoretically, and in the mid-1990s, computational power and memory weren't up to it anyway. Marder also knew that Abbott would resist the idea of working on such a realistic and detailed model. "Larry would never have touched it. It was implausible. But," she told me in 2014, "that was always my dream, and Tim O'Leary, a very clever postdoc, originally a mathematician, is getting there now. But we have a completely different level of computing power now. And we know a lot more biologically."

In the models devised in Marder and Abbott's lab, calcium was stipulated as an important component of the sensor set point, the feedback mechanism. When a neuron's level of activity rises, say from tonic firing to bursting, certain ion channels in the neuron's membrane that are sensitive to voltage open to allow an influx of calcium ions. Calcium ion concentration inside the cell is increased; for the models, these levels were taken as indicators of a cell's recent activity. But intracellular calcium is a major player in biochemical pathways that can lead to a variety of results, including—and most relevant to the regulation of a neuron's excitability—the transcription of genes coding for the proteins that assemble to form ion channels and receptors. Calcium ion concentration, increased by the electrical activity of the neuron, can thus affect the synthesis and placing in the membrane of ion channels and receptors, and this change in the number of channels and receptors can therefore be said to be activity-dependent. Moreover, calcium ion currents are important in synaptic transmission and electrical excitability in general. Calcium is involved in so many cell processes that it is difficult to isolate and study just one of them. Hence, to prove experimentally that the modeling assumption was correct, calcium ion concentration levels would have to be manipulated in a biological neuron. Of course, it's a vicious circle: if calcium ion concentration is manipulated, then the usual "reporters," such as the membrane potential, are disrupted too. It was what is justifiably called an embedded intellectual problem, and it tormented Marder. "I didn't see a way in. Any time I came up with a clever experiment, any manipulation I could think of doing would interfere with the output measure. I always got stuck. I'd think it through, and I'd end up knowing I didn't have a reporter any more. If you disrupt something that you think is part of the central regulatory process, but your report on what's going on depends on that, you can't break the cycle. I got very fed up with everybody saying," and she imitates a fussy voice, "Well, can you prove it's really calcium?" She sighs and shrugs, "And I'd say, 'Yeah, if I mess up with calcium then I don't have a burst do I?' and they'd look at me and say, 'Oh, right, yeah.'"

In the first six months of 1994, Marder was able to take a sabbatical. She chose to spend the semester in Mary Kennedy's lab at Caltech in Los Angeles. Kennedy was the preeminent expert on the biochemistry of a molecule called calcium-calmodulin-kinase-II, and Marder had a hunch that if calcium-binding proteins were part of the homeostatic feedback mechanism she wanted to explain, it would be the CaM-kinases, as they are known for short. She was wondering whether CaM-K II was a link in the regulation pathway. Today, she says she was probably more convinced of it then than she is now.

Sabbaticals are supposed to provide a change of mental scene and allow time to review one's work. I asked Marder whether the sabbatical had given her any new perspectives. She laughed: "No. I've never found sabbaticals to be terribly enlightening or illuminating. It just means you tell everybody you can't do their committees because you're on sabbatical. And it takes them all a little while to find you again, after you've come back. That's the important thing."

You can also, as her team of 1994 teasingly points out, manage to miss the major upheaval of moving the lab into a new building. "She was in sunny LA," they pretend to grind their teeth. It was a bigger lab and purpose-built. "There was a wet lab with Eve's office at one end, and Larry's at the other. It got drier toward Larry," says Frances Skinner, who had a dry office. Was she consulted about the design, I asked Marder. "Oh yeah, it was a pain in the neck." Perhaps, but her insistence on having a common room with a kitchenette for her lab was worth the pain—it is the team's social center, its lunchroom, its debating chamber, and the weekly rendezvous for presenting work semiformally to each other.

But when Marder came back from her sabbatical, she was still stuck. She was sure that nothing could be done in lobster cells, but there were better tools for investigation in mammalian cells. She could imagine making new CaM-kinases, for example, or manipulating signal transduction pathways, and she even says that if there had been a simple experiment in mammalian cells, she might have considered doing it herself. It's still one of the lasting problems that exercises her mind, although, almost twenty five years later, she can say, "We now know enough about a myriad of molecular mechanisms in different kinds of homeostatic regulation to say that neurons use not just one but multiple processes to couple robust network performance to the animal's history because that is so critical to neuronal and network stability." But, in 1994, she talked the problem over at length with Turrigiano and finally told her, "Gina, I really hate this, but I feel the problem needs to be moved to mammalian cells; they have reagents, a lot more well-defined pharmacology, the genetics ... I'm probably not going to do it, but you might want to."

Turrigiano did make the switch to mammalian cells after leaving Marder's lab. Because she was working with dissociated cells at the time, the transition wasn't as big a step as it might seem. Her move wasn't a big step either, as she got a faculty position at Brandeis. However, in 1994, Turrigiano's track record was entirely associated with the lobster, and that would have made it difficult for her to get funding for work on mammalian models. On the other hand, her work on homeostasis was significant and readily fundable. So, allowing Turrigiano to apply for funding on that basis, Marder, in one of her grander generous gestures, said that she herself would stop working on homeostasis in lobster stomatogastric ganglion cells until Turrigiano was well established in her own lab. Three or four years later, when it was clear that the Turrigiano lab had taken off, having made a successful transition to rodent neurons, Marder picked up the work again.

The other crucial unresolved problem in demonstrating neuronal homeostasis is that so far no satisfactory model of it exists at the network level. In the second half of the decade, Marder's attention was drawn to what seemed to be a promising experimental approach. Evidence from other labs showed that a preparation of the ganglion, cut off from the commissural and esophageal ganglia and the stomatogastric nerve and thus deprived of all modulatory input, would slow down or go silent. That phenomenon was already well known, and it chimed with Marder's own experience in Paris, where she had struggled with the unresponsiveness of ganglia prepared without the front end. It was thought that loss of activity, or loss of the modulatory input that promotes that activity, silenced the neurons. But now researchers had noticed that, after a few days of isolation in a Petri dish, the ganglion seemed to revive and exhibited rhythmic bursts again. This phenomenon was being called "recovery of function." It seemed like a good way into the network homeostasis problem, something that could perhaps be followed up, unlike the calcium conundrum.

Jorge Golowasch was again working in Marder's lab, and he discovered that, after a day or two of inactivity, an isolated ganglion seemed to start up in distinct bouts of activity. Then, a couple of days later, it would settle into rhythmic activity similar to its normal workings—similar but usually notably slower. They speculated that perhaps this slow rhythm represented the ganglion's activity in the most usual state of the animal—unfed and therefore making no call on the digestive system. In computer models of the phenomenon, the model neurons' conductance densities changed during the recovery process and stabilized at levels that were different from those in the ganglion, although still efficient.

But Marder was uneasy; the preparation behaved erratically in both her own lab and other people's. Sometimes the ganglion would be silent within about twenty minutes after being isolated from all input; in others it kept going quite well for days. Over the next few years, other stomatogastric ganglion labs were also troubled by their unreliable experiments. The main problem was that, after the early rush of interesting results over a couple of years, the ganglia stopped stopping. Or not. Marder sighs: "I can ring round some stomatogastric labs and ask, 'Are they stopping now?' and people will tell me they're stopping but they want them to keep going, or vice versa. Whatever you want them to do, they do the other thing."

Marder, as usual, looked for clues in the natural life of her crustaceans. She thought that the underlying state of the network, reflecting the animals' recent history, might be one of the reasons. Was the phenomenon seasonal? Was the temperature of the ocean when the crabs were caught important? Was it hormonal? This time, the wild crustaceans have kept their secrets. Of course, these questions would not have to be asked in work with lab-raised animals.

Or was the recovery phenomenon itself an artifact? Recently, more than a decade after the initial flurry of interest, a doctoral student in Marders' lab decided to record continuously from preparations over many days. When recovery of function was first investigated, this would have been impossible because the amounts of data generated would have overwhelmed a lab's computers. Every now and then, Golowasch would record an experiment for 48 hours continuously, and that was considered noteworthy. But by 2013, Al Hamood could record for weeks, a good example of new possibilities brought about by better technology.

Marder insisted that Hamood set up a continuous perfusion system because one of the possible confounds in earlier work, along with the limited recording sessions, was the abrupt changes of the saline bath the preparation was in. Researchers would take the dish off the rig and put it in an incubator to keep overnight. The next day, another session would start with putting the dish back on the rig and renewing the saline. Hamood guessed that the saline changes could have reactivated the ganglia so that what researchers were attributing to recovery of function was actually an artifact of the method. Remember that what everyone familiarly calls saline is not merely salt water; it is actually a savant recipe for each species, including sodium chloride, of course, but also salts of potassium, magnesium, and calcium, Tris base (2-amino 2-hydroxymethyl 1,3-proanediol), maleic acid, and even dextrose for *Panulirus interruptus*. Hamood's data showed that with intermittent complete saline changes there are reactivation periods, whereas with continuous saline perfusion the ganglion only infrequently goes silent in the first place.

The idea of studying this recovery phenomenon as a manifestation of homeostatic regulation at the network level was seductive. The statement of principle that was on everybody's lips would have tied homeostatic regulation of individual neuronal excitability tidily into the network: the network would be stabilized by activity-dependent regulation of conductances in participating individual neurons. There would be two operational states of the ganglion that produced its rhythmic activity: either dependent on the neuromodulatory inputs from the anterior ganglia and the stomatogastric nerve or emerging from the ability of the individual neurons to respond to perturbation by adjusting their excitability.

But Marder says, "I just don't know what's going on. The data—wrong or right—were what we got at the time, and we interpreted them as we did.

But I don't have my finger on it. It's clear from data in several labs that changes in channel expression do occur over time, and that these could contribute to functional recovery. But the phenomenon remains somewhat capricious." With hindsight, taking later work on variability into consideration, she adds, "Perhaps capricious is the answer: depending on each animal's initial conditions its ability to recover partially or fully from the perturbation will be different."

It is rare for Marder to drop a line of research; apart from this recovery of function model or the work on development, her major themes continue to be studied throughout the years. So I asked her whether she had other candidate phenomena for investigating network-level homeostasis. She said there were other ways of doing much the same thing—isolating the ganglion—without the confounds of salines and lab protocols. For instance, a graduate student in David Barker's lab had cut the stomatogastric nerve in living animals. These large crustaceans can, and often do, survive long periods without much food, so there was no immediate effect on their health. After several months she checked the animals to see what the ganglion was doing, and there was still some activity. But, of course, the ganglion was still bathed in hemolymph and getting hormonal input. Marder says, without enthusiasm, "We could try something like that. But the essential problem remains: network output depends on too many things. If I go back to the recovery model, it'll be at the single-cell and single-molecule level. Probably I'll just sit and wait. There are happy endings, sometimes."

* * *

Whatever the frustrations in finding a good experimental model of network homeostasis, Marder's intellectual attention was fully engaged in examining the principle of homeostatic stability. She would now take the paradigm a step further. But first let's look at a short summary of how she advanced her theory, analyzing the contrasting attributes of stability and plasticity.

When Marder started her research, investigation of network function was principally focused on connectivity diagrams, showing synaptic relationships between the neurons of a circuit. She soon saw that the wiring diagram of the stomatogastric ganglion's circuits was not the whole story but even so, too complicated to exploit using the techniques then available. The diagrams indicated only excitatory, inhibitory, or electrical connections, whereas Marder wanted to describe the connections in greater detail, starting with the chemical messages sent and received by each neuron. Her view was that a single neuron might well signal using only one substance,

but it was likely to receive more varied signals conveyed by different neurotransmitters. As we have seen, this discerning line of thought led Marder
to recognize the importance of the neuromodulatory substances that influenced each neuron's state and its constantly changing environment. She
began to study the responses of neurons to a wide variety of influences not
shown on a wiring diagram at all.

The diagrams looked static, but Marder discovered a surprising level of
plasticity. Her experiments led her to realize that neuromodulators could
produce effects on the ganglion so strong that the accepted borderlines
between its circuits could blur and change. Studying the multifunctional
networks and the neuromodulatory conditions that reconfigure them,
she was increasingly concerned with the question of control: how was
the amplitude of a network's response to neuromodulation controlled?
What were the mechanisms ensuring that the network stayed within its
operational range? To be useful to the animal, the behavior of networks of
neurons must be stable, but they are subject to a great many modulatory
influences all at once. At least in experimental conditions, networks could
be significantly perturbed in this way. Yet the wild lobster gets on with its
life for years in all sorts of conditions that must affect the modulatory influences reaching its stomatogastric ganglion. The issue of stability became a
constant undercurrent in Marder's thought.

Neuroscience has long wrestled with these two contradictory themes—
plasticity and stability. The apparent paradox is perhaps most pronounced
in the field of research that looks for the workings of learning and memory.
Any such mechanism must operate through the "plasticity" of neurons and
networks because clearly some sort of change is involved when an event
is experienced or something is learned. But for memory to persist, that
change must be stabilized.

Most learning and memory research has been focused on what happens
at synapses, on activity-dependent changes in the strength of these connections. Note that the word "synapse" can be used to mean the simple
fact that one neuron connects to another (neuron A synapses on neuron
B) but also to mean the closely apposed structures in the cell membranes
of two neurons. In this latter sense, the two neurons can have many, many
synapses, and the number and strength of these synapses can change.
In 1949, Donald Hebb introduced the idea that when two neurons were
strongly active simultaneously and their activity was correlated, that activity could produce long-lasting change to their common synapses. Hence,
the mantra known to all neuroscientists is "neurons that fire together
wire together."

As in memory formation, so in other processes in living organisms; there must be sufficient flexibility to meet the different challenges of a lifetime but also sufficient stability to ensure the organism's continuing functions. Marder's prolonged reflections on the changeable electrical properties of neurons brought into focus and defined the fundamental question: how is stable network function maintained throughout an animal's life while allowing growth, learning, neuromodulation, and appropriate responses to environmental perturbation?

Her work with Abbott was pointing to an answer: homeostatic regulation of neuronal excitability, at least under conditions of simulated or experimentally imposed perturbation and at least at the level of the single neuron. The discovery that neurons have a lifelong identity in the kinds of ion channels they express represents a form of stability. Conversely, Marder's work with Abbott and Gwendal LeMasson had shown that neurons have no set numbers of the ion channels that give rise to their electrical properties. They respond and adjust to recent activity. They are plastic. The observation that neurons adjust after a perturbation, responding to feedback about recent activity, and then are able to fulfill the same function demonstrated homeostasis. Tellingly, the concept of homeostatic regulation brought the two attributes together and signified that plasticity is necessary for stability.

Now Marder took the idea of activity-dependent regulation further, indeed out of the laboratory altogether, by proposing that it was constantly at work throughout an animal's lifetime. Why wouldn't homeostasis be a continuous process in the daily life of the lobster or the lion? Was there any reason why homeostatic adjustments should be triggered only by significant perturbation? That would imply a threshold of disturbance at which neurons started to respond to activity-dependent feedback. It is difficult to imagine what such a threshold would be. True, homeostatic regulation had been revealed by experiments and models using significant perturbations, but in the living animal neurons are affected all the time by naturally occurring fluctuations of neuromodulation. Her conjecture was that, as an underlying mechanism ensuring stability, homeostatic regulation must be ongoing at all times. When a neuron responded to natural neuromodulatory influences by raising or lowering its level of activity, however slightly, the same sort of feedback mechanism would cause continuous fine adjustments, allowing the neuron to stay within its functional range. Marder called it "tuning to target."

No one had been able to see it experimentally. That would certainly have been impossible in the first half of the 1990s, and even today it would be hard to do. Musing about this, Marder imagines, "If you could magically image calcium continuously for days and days in response to either minor fluctuations or major fluctuations, you should actually see what's going on—if the calcium is in fact a major player in that regulation. But in any case, most people who image calcium do not image it quantitatively. We can't do that continuously for many days on end."

It had been assumed for decades that, just as a neuron could be recognized by its anatomical position, morphology, transmitter use, and characteristic firing pattern, so too could its conductances be measured and were part of that cell's identity. Those conductances corresponded to populations of specific ion channels, and, thus equipped, the neuron could fulfill its role. Now Marder and Abbott were saying the opposite: the neuron's ion channel populations are constantly adjusted so that the neuron can do its job. The necessary function or target activity elicits adequate electrical properties in the membrane at any one time.

Tuning to target was a radically fresh formulation based on three new, clearly articulated principles. First, the neuron's role in the circuit, the activity the circuit requires of it, is the neuron's target. A neuron uses feedback from its recent activity to regulate its electrical properties, thus enabling it to conserve its characteristic pattern of activity. Adjustments are made so as to maintain the required activity under all the varying circumstances of the animal, not so as to maintain a particular number of ion channels that are part of the neuron's identity.

Second, each neuron's population of ion channels is adjusted, not one type at a time according to some functionally desirable number for each, but according to the composition of the whole population, the balance of strengths of the different currents. Therefore, no "correct" or canonical values, valid at all times, are to be found for the conductance corresponding to a given type of ion channel in a given type of neuron.

Third, the rule for tuning must rely on feedback from some sensor in the neuron that activates the pathways enabling the adjustments. The neuron must have a core mechanism tending to restore it to an efficient state—a homeostatic mechanism.

Tuning to target is a strong and fundamental concept. In an exchange typical of Marder, I asked her whether, when she started to talk about it, she had a feeling she was way ahead of the pack. She assented but then dodged any hint of competitiveness, saying, "I just thought it was really cool." In a serious vein, I asked Marder what persuades the neuron to this constancy. She answered, "I think all the molecular events leading to determination of cell identity are involved in setting up the molecular configuration of the transcription factor pathways required for these adjustments. The implicit assumption is that the set point of the target for homeostasis that lives in each neuron has to be part of the determination of cell identity."

Here, Marder is referring to a decisive step in the progression from the apparently identical cells of an early stage embryo to specialized cell types. Determination involves such processes as activation of specific genes and responses to extracellular signaling molecules. The development potential of each cell is increasingly restricted as the cell commits to its "fate." Fate in this context means the type of cell that the progenitor will become: a skin cell, a fat cell, a pyloric dilator neuron, and so on.

Unlike most other cells in the body, neurons last for the lifetime of an animal, which may mean days, weeks, years, or decades. Even here, of course, any idea of lifetime permanence must be tempered. Marder has always been fascinated by the ability of neurons to carry on, showing up for work, as it were, while receiving so many disturbing influences, and with the neuron's structures constantly turning over. She often likens the neuron to an airplane flying on course at thirty thousand feet while every nut, bolt, fin, flange, and cable is being replaced with parts made in the plane as it flies.

It's an apt metaphor. Ion channels are assemblages of proteins that span the cell's membrane; they allow the small charged particles called ions, those specific to each channel, into or out of the cell. An ion channel has a lifetime measured in hours, days, or weeks. Receptors have a lifetime often measured in days or weeks and can be wondrously complicated. Some are long threads of peptides that weave in and out of the membrane according to the affinity or aversion to water of different parts of the thread. (Inside the lipid membrane, molecules are sequestered away from water, whereas both the cytosol inside the cell and the extracellular milieu are aqueous.) These extraordinary compositions are synthesized unceasingly, in varying numbers according to the cell's activity—along with all the rest of the cell's components. You have to imagine, inside the neuron, a seething crowd of ions and molecules. The molecules are assembling into large proteins, folding into complicated sheets and spirals, then joining up with other assemblages, and constantly shifting around. Some structures move into the membrane, others form firmer connections with each other and scaffold the cell's shape. It's all a far cry from the simplified and firmly outlined diagrams we use in our efforts to understand.

This brings us back to the question of experimental timescales. Obviously, any process that relies on the synthesis of proteins will be on a longer timescale than the neuron's immediate responses to many experimental interventions. The proteins that are components of ion channels and receptors must first be called into the fray. The genes that allow synthesis of the proteins are contained in the cell's nucleus. Those genes must be activated by messengers responding to the cell's requirements; one of calcium's many roles is in this sort of pathway. The proteins are produced and assembled in stages on their way to the particular area of membrane where they are to be inserted. At the same time, "old" receptors and ion channels are being withdrawn from the membrane, internalized into the cell, dismantled, and disposed of. A staggering number of steps are hidden behind the simple phrase "tuning to target."

* * *

If homeostasis helped to explain stability, then what mechanisms underlay plasticity? Toward the end of the 1990s, Marder explored the question. Work in her lab would now help to clarify this part of the dichotomy, and Marder started to develop the two complementary concepts that came to be called "variability" and "multiple solutions."

In 1997, Marder and Abbott, with Abbott's student, Zheng Liu, embarked on an ambitious new model. They overcame some of the limitations in the earlier models that were due, for example, to the use of a single calcium ion feedback sensor. Now they included three: a fast sensor, a slow sensor, and a direct current filter that tracked the inward calcium ion current. Each model neuron was regulated by a programmed mix of signals from the three sensors. Thus, the maximum conductances were not controlled using just one sensor, as in earlier models, but depended on this balance of feedback.

The first homeostatic model neuron, four years earlier, had been programmed to return to a single equilibrium point, a specified activity pattern. As Marder remarks today, "That was fine then because everybody was still living in a world where cells would have a canonical set of conductances and any noise we saw we just assumed was experimental, or whatever—I don't think we actually questioned whether there was a single solution, so that model was structured to give you a single solution. Also, the neurons that Gina was looking at in culture were all moving in the same direction to what looked like a target activity pattern."

This second homeostatic model was less constrained than the first; it stipulated a particular activity pattern as the end point, or target, but during the run, the strengths of the different conductances were no longer specified at all. The starting point for each run was a random set of maximal conductances based on data from Gina Turrigiano's experiments with isolated neurons. When a simulated perturbation was introduced, the model neuron would adjust its properties to revert to the target activity. When the perturbation ended, the model neuron went through further adjustments that brought it back to that activity pattern but not necessarily with the same conductance values as at the start of the run. Running the model a thousand times, the model neuron reached the required bursting pattern in about ninety percent of them. What astonished them all was the range of maximal conductances the model neurons could use to arrive at the same activity. Could this theoretical model be right? Marder realized that the experimental data on the lateral pyloric neuron's conductances that Jorge Golowasch had collected for his doctoral work in the late 1980s might show something like it and grimly remembered the week in Paris. "Because when we were writing his papers, we were constantly worrying about whether to use best data or median, so I knew there was a spread of data there. And I had been so scarred by that. So the only question was whether the data were recoverable."

Golowasch went to look: "I had never looked at the data that way and started seeing that there was a lot of variability in no matter what current I had measured." It was a big surprise: depending on the specific type of channel, the maximal conductance measured varied from two- to fivefold between preparations. At the time, intent on describing each of the conductances with its amplitude, time course, voltage sensitivity, and so on, the values for each measure in several LP cells were simply averaged; the variability was taken as a normal range for experimental work and didn't prompt any particular reflection. Golowasch and Marder did note the balanced relationship that prevailed between the maximum conductances for the outward currents thus described but did not think of unbundling the data. This time, some ten years late, they did plot the range and correlations in the data. The range was comparable to that in the model, although the absolute magnitude of conductances in the biological cell was much smaller. This difference in magnitude was to be expected. Golowasch's data and the data used in the model, Turrigiano's, had been collected in different experimental conditions. Also, the biological LP and the model neuron had different activity patterns: the LP would have been in tonic firing mode in the voltage clamp experiments, whereas the model neuron was instructed to reach bursting activity.

The new examination of Golowasch's results was included in the paper describing the second homeostatic model, published in 1998. "A Model Neuron with Activity-Dependent Conductances Regulated by Multiple Calcium Sensors" was a turning point in their thinking and explicitly introduced the concept of variability, both animal to animal and moment to moment. The discussion section argued that, because the wide range of conductance values shown by the model had also been found experimentally, " an identified neuron displaying a stereotyped activity pattern, such as the LP neuron, could have significantly different sets of conductances when measured in two animals or in the same animal at two different times. Perhaps some of the variability in physiological measurements of membrane currents that has been attributed to experimental error may reflect instead an intrinsic variability inherent even in identified cell types."

The idea hinted at a wealth of information missed in analyzing data, an uncomfortable thought for biologists. I asked Marder what reactions to the publication she had from her peers, and she said, "The real theorists knew this already. Experimentalists were still dubious at that point—sort of nervous. It took them a long time." Marder could easily accept the idea of variability. "That was fine. Because we were doing the homeostasis stuff we assumed that moment to moment and day to day the numbers of ion channels in cells would be bouncing around. So when Jorge was seeing this variability we didn't necessarily find it inconsistent." Moreover, one shouldn't forget that this possibility had occurred to her many years earlier when she was writing her thesis. In a discussion of the experimental difficulties she had encountered, she had written: "The variability in the reported measurements may have been a result of the combined physiological identification and dissection treatment or may reflect large actual differences in the activities in the cells."

Marder and Golowasch wanted to be sure that variability was a real phenomenon, not measurement error in their own work, and to try to settle the question of its cause: was this variability due to different properties inherent in each animal's neurons, or was it caused by the dynamics of inherently similar neurons reacting to different recent activity? Having wrestled with the choice of data in their earlier work on the LP, they were very much aware of the importance of gathering consistently comparable data that would be unassailable in this new and sensitive context. Golowasch went to work on new voltage clamp recordings using the inferior cardiac neuron (IC), of which there is just one in each stomatogastric ganglion. He chose to measure three outward potassium ion currents because they were the easiest to measure accurately, and there were pharmacological agents and membrane potential manipulations that allowed each of these currents to be unambiguously identified. These currents are present in all types of stomatogastric ganglion neurons.

Golowasch gathered data from eighteen preparations that revealed two-to fivefold differences between animals in the three currents. Because this range matched that found earlier and shown by the model, this wide range of values could now be considered valid; it did not indicate inaccuracies or experimental errors. Most intriguing was the fact that the evidence, both theoretical and experimental, showed that this significant range of variability was no bar to achieving the typical patterns of activity necessary for the animal's behavior. Similar output could be obtained from neurons of the same class, both model and biological, that had quite widely different combinations of conductance densities. There seemed to be no canonical values for conductances in neurons of a particular type, just a functional range. Moreover, the balance between the currents was also different in each animal (see figure 8.2a).

Next, Golowasch tested the influence of recent activity by giving each cell three hours of pulsed inhibitory stimulation (similar to its normal conditions in the ganglion). All the cells now had the same "history" to which all responded within the first hour, slowly changing their conductances in similar ways—and then slowly reversed the changes after the stimulus ended (see figure 8.2b).

The results suggested that the variability in conductances recorded in many measurements might be the result of each neuron's recent history, in fact reflecting different prior states of activity of the stomatogastric ganglion.

Talking with Marder today about the work that led to her focus on variability, she frames her story as usual, in terms of her collaborators' work; it is quite difficult to persuade her to talk about her own contribution. Finally, I pushed Marder harder for an answer, and she replied, "Where I probably played the strongest role is, I probably gave Jorge permission to take his data at face value. I think up until that point, every biologist would have been busy averaging things and just saying, you know, all of this other stuff is experimental error and let's push it under the rug. Especially in voltage clamp, when you get variability it makes you nervous. And probably I was the person who was able to step back and say, well, you know, he's been doing this for enough years, and he's good enough at it—he's got very good hands—and he's generating this data, then we have to believe it."

Marder's confidence was all the stronger for knowing that Golowasch had not even been looking for variability when he got the 1980s data. He was measuring in voltage clamp, passing currents to perturb the neuron's activity and asking whether that would alter the balance of potassium conductances. The important point was that, in so doing, he collected voltage

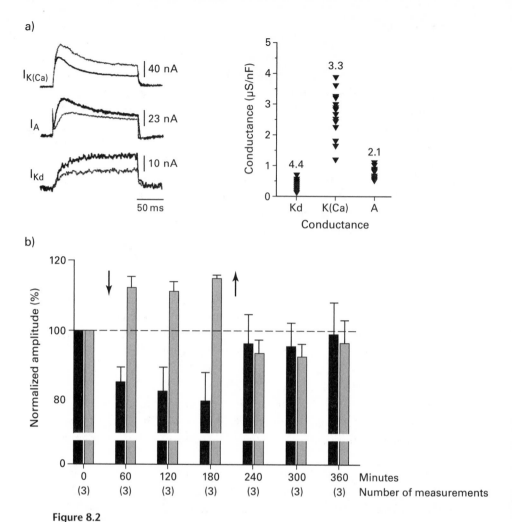

Figure 8.2

(a) Left-hand panel: The three outward potassium currents measured under the same conditions in two inferior cardiac neurons (grey and black traces). The neuron with greater $I_{K(Ca)}$ current amplitude (grey traces) has the lesser amplitude for the two other currents, indicating that no predetermined balance exists between the currents in these neurons. $I_{K(Ca)}$ is the current activated by calcium ion influx, I_A is the rapidly in-activating current that responds to the influx of sodium ions that leads to an action potentials, I_{Kd} is a delayed rectifier current that controls the duration of action potentials. Right-hand panel: Peak conductance density levels of the same three outward potassium currents measured in eighteen inferior cardiac neurons. Each point represents measurements from a different preparation. The figure above each column indicates the ratio of maximum to minimum values. (b) Histogram showing maximum current amplitudes as they change during stimulation and return to previous values over a similar time period. Stimulation of 3 neurons over three hours started at the downward arrow, followed by three hours without stimulation after the upward arrow. Black bars: sum of I_{Kd} and $I_{K(Ca)}$. Grey bars: I_A. (From Golowasch et al., 1999.)

clamp data on multiple currents in the same cell, which was something that people would not normally have done.

Marder added, "The models had shown that you could have very similar behavior with different underlying structures so, to my mind, it was the models that gave us permission to trust our data. I had the confidence and the authority to pull that off and I knew that the models were telling us that it could, had to be, this way. And therefore we should trust the data, or at least take it seriously. Because people have been getting variable data since the cows came home, right?"

In Zheng Liu's model, the neuron achieved the requisite activity pattern in ninety percent of the runs, but it got there by using many differently balanced sets of maximal conductances—many different solutions. Looking around, Marder saw that other people's models seemed to work that way too: "You would find indications of that in the literature and it was clear to me that it was a much better way to run a biological railroad because then you didn't have to get everything so carefully tuned. You had tolerance." Marder began to formulate the concept of "multiple solutions."

* * *

In 2000, a newcomer turned the lab's modeling logic on its head. Mark Goldman was a graduate student with Larry Abbott. Taking his cues from the Liu model and his data from Turrigiano's work, he decided to build a new type of model. Instead of trying to get a good representation of a certain phenomenon by specifying a target activity for the model neurons, his model let the activity patterns emerge when five maximal conductances were randomly run through a range of values. This generated thousands of conductance sets—thousands of model neurons. The resulting activity patterns could be characterized as silent, firing tonically, or bursting, with various numbers of spikes per burst.

Marder always speaks of Goldman's model as pivotal in her understanding of multiple solutions. Of course, if you set out to look at thousands of conductance sets that bring about a few target activities, by implication you are looking for multiple solutions. In fact, the report on the model, published in 2001, is not ostensibly about multiple solutions and doesn't mention the term. Catchily titled "Global Structure, Robustness, and Modulation of Neuronal Models," it examines the stability (robustness) of neuronal responses to variations in neuromodulatory input (represented by the conductance changes such input would cause). The thousands of model neurons did show stability in the sense that, although their varied properties were distributed in a big cloud of parameter space, most of them still brought about some characteristic neuronal pattern of activity. However, most of the paper is devoted to Goldman's finding that certain types of changes to the "parameter space" (which encompasses all the values generated by all combinations of all the parameters in the model) were more effective than others. In the real world, they would have resulted in greater sensitivity to neuromodulation. Jorge Golowasch confirmed the predictions of the model in stomatogastric neurons from the crab, using an improved dynamic clamp program that had been developed by two of Marder's postdocs, Yair Manor and Farzan Nadim.

This filled out the lab's understanding of why modulators worked in some states but not in others, and how some neuromodulators primed a neuron for other modulators without themselves affecting that neuron in any detectable way, or why some neuromodulators affected a neuron's conductances only slightly but produced big changes in its intrinsic excitability. Again the conundrum had to be explained: if neurons could be so variously sensitive to neuromodulation, how could they remain stable?

Yet what Marder retains from the work is something different. "That paper really had all the seeds of multiple solutions in it. I think it was front and center in Mark's mind. That may not come through because I don't think I understood it as well when we were writing it as I do now. But if you look back with hindsight, you can see it all there." For her, two of the figures in the paper were unforgettable: one showed identical "behavior" in model neurons with quite different conductance sets and the other, conversely, showed similar conductance sets resulting in different patterns of activity (see figure 8.3).

While Goldman was engaged on this project, he noticed a curious and troubling phenomenon in the operation of averages. It was a chance observation, mentioned almost in passing in the report, but it led to an important insight. Goldman devised a way to test his results by using a subset of his model's simulations to look for a single type of activity: one-spike bursting. He ran 2,000 "neurons," meaning 2,000 different conductance sets. There were 164 one-spike bursters. Then he took the averages of each of the conductances in those sets and used them to form a single set. The set of conductances—the model neuron—that was formed astounded them all: it turned out to be a three-spike burster, unlike any of the 164 model neurons it was supposed to represent. This meant that when considering more than one variable in a set of nonlinear relationships, the usual statistical reflex of the biologist—to calculate the means and standard deviation—could result in an average that did not correspond to any of the population in the study.

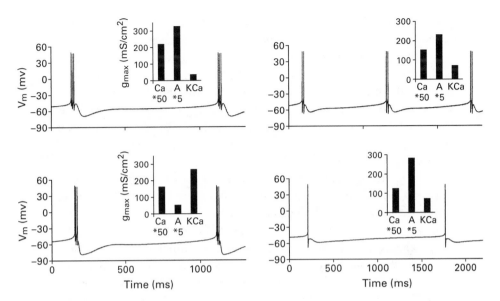

Figure 8.3
On the left, identical three-spike burst wave forms that were produced by two model neurons with different conductance sets (insets). On the right, very different activity, bursting and tonic firing, results from only slight variation in conductances. Ca conductance multiplied by 50; A conductance multiplied by 5. (From Goldman et al., 2001.)

The finding was published in 2002 with the title "Failure of Averaging in the Construction of a Conductance-based Neuron Model." Golowasch says the lab was aware that these discoveries were radical, but he acknowledges that it took a long time for them to be more widely appreciated. "The averaging paper was in the *The Journal of Neurophysiology*, and the global robustness one was in the *The Journal of Neuroscience*—both top journals—but these earth-shattering ideas aren't recognized unless they come in a flashy journal like *Science* or *Nature*."

It was certainly not a trifling matter; it meant that computer models using averaged data might be sound—or might not. Unfortunately, no one can tell when a failure of averaging is likely because no one knows precisely what the characteristics are of data that would produce one. No one could judge, from inspection of experimental data, which individual conductances had key relationships with other conductances for producing a behavior.

* * *

Marder calls her intellectual progress a straight line. After her first research on neurotransmitters, her conviction that the existence of so many different ones must have a functional purpose led her to ask how the dynamics of circuits are organized. "Then neuromodulation, circuit reconfiguration—circuit switching just comes out of reconfiguration—and then at some point you have the branch into trying to build computational models, which leads immediately to homeostasis. And then homeostasis segues into variability and multiple solutions. With neuromodulation running constantly throughout."

Her straight line now ran from Zheng Liu's work, the second homeostatic model, to the work described in the next chapter that clarifies and confirms the twin concepts of multiple solutions and variability. Multiple solutions can only exist if there is variability in at least some of the underlying values. That variability can emerge from homeostatic processes in which feedback from recent activity affects intrinsic electrical properties. Because a neuron constantly tunes to its target activity, today's properties could be different from yesterday's or from those of another animal. Tuning to target therefore drives variability, and that variability entails multiple solutions as each animal must maintain its vital behaviors using whatever properties it has at that instant in time. As Marder says, "Those properties just have to be good enough."

Marder thinks that theoretical work forced her to examine her assumptions and identify the gaps in the logic of her "word models"—to be more disciplined in fact. But on the other hand, she has called modeling "legitimized dreaming" because speculative models suggest new explanations of laboratory results and new avenues to investigate at the bench. Whatever the contradiction may be between these views, in a decade, Marder's thinking and her work with Abbott and their labs had upset or reversed five accepted paradigms. The new versions are listed succinctly below, summarizing the suite of ideas.

A specific type of neuron does not have canonical electrical properties that can be described once and for all and printed in a textbook.

The electrical properties a neuron exhibits at any one time are not necessarily constant for that neuron; the balance of its conductances is likely to vary, reflecting the neuron's recent activity, while remaining adequate to fulfill the neuron's function. Thus, there is variation animal to animal as well as between neurons of the same type in the same animal.

A neuron's excitability is not the governing factor for its activity pattern; its recent history of activity regulates its excitability.

No one exact balance of conductances is required to produce a particular pattern of activity; many sets of conductances can lead to the same activity pattern.

The range of values for experimental data that should be treated as valid is likely to be wide because the variability between one animal and another is significant. Outlying values in data sets that have habitually been treated as "noise," the result of experimental error, and omitted from reported results and statistics, are likely, instead, to be informative.

Marder now had at least part of an answer to the profound question of how neurons function adequately through development, growth, and the turnover of constituent parts. She could describe how a level of stability good enough for neuronal and network function under changing conditions might be achieved. However, it was clear that the consequences of these new lines of thought, their implications for neuroscience, were far-reaching and difficult to deal with. Marder admits that a 2009 research report described in the next chapter, "Functional Consequences of Animal-to-Animal Variability in Circuit Parameters," was "the most despairing paper" ever to come out of her lab.

9 Good Enough

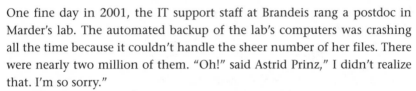

One fine day in 2001, the IT support staff at Brandeis rang a postdoc in Marder's lab. The automated backup of the lab's computers was crashing all the time because it couldn't handle the sheer number of her files. There were nearly two million of them. "Oh!" said Astrid Prinz," I didn't realize that. I'm so sorry."

Prinz had been simulating 1,679,616 different sets of conductances. She had broken the simulations into chunks and loaded them on all the lab computers to run in the background, using only idle processor time. Naturally, that meant most of the simulations ran at night. Every morning, Prinz gathered her data. It all made sense, except that Prinz was not a computer scientist, and she had made a rather elementary mistake: she had given each conductance set its own output file.

There are two versions of the background to this story. Marder says that Prinz wanted to have a good model of the PD cell, "So she started with Zheng Liu's model and she hand-tuned for about a week. I remember going by her desk and she was cursing ferociously in German. Then she went silent for a while, and then I discovered that for several weeks she had been running these multiple models in the background of every computer in the lab. At some point, to avoid the hand tuning, she just decided to put everything on a grid, and she picked six values for each of the eight currents and then decided to simulate all against all. When she eventually told me, I said, 'Hell!' and told her to get them off all the computers and onto the cluster."

But Prinz clings to a different and more likely series of events. She says she was trying to do something else entirely, but when she had done it, Marder turned her story into gold. Nevertheless, they both agree that Prinz started on the project without detailed consultation, somewhat naively thinking it wouldn't take much time to set it up.

Prinz had been working on phase response curves—the magnitude of an oscillatory neuron's response to a perturbation depends on the stage of its cycle period at which the perturbation occurs. She remembers a conversation in Marder's office, discussing what properties of a neuron determine its phase response curve. "And we both said we don't really know. Then I remember Eve said this one particular thing: 'Well maybe at some time we should do a parameter space exploration on that.' Mark Goldman had done some parameter space exploring, but I think the timing was good for me in that the progress in computing power and what was feasible computationally even just from a couple of years before when Mark had run his simulations was far, far better. We could really consider large scale explorations. But my motivation was: we just don't know what these conductances do and so let's explore all of them. It wasn't a novel idea."

It was, however, a novel approach: a big database, number-crunching model. Two months went by in writing the code for a single compartment model that used eight conductances taken from Gina Turrigiano's work. For each conductance, there were six evenly spaced values from zero to the maximum Turrigiano had found. That meant 6^8 combinations; each combination is a different model neuron, so there were 1,679,616 of them. The program had to be efficient—clearly, Prinz had to minimize the simulation time per neuron as much as possible, so she made the model move on if a combination "did something boring." Her limit was five cycles if they were purely repetitive, but some model neurons did not fall into regular behavior that quickly, and the simulation then had to be run for much longer to understand why. There were months of tryouts and fiddling with algorithms.

When Prinz had scanned the whole lot, she found that, of the 1,679,616 model neurons, approximately seventeen percent were silent, sixteen percent fired single spikes regularly (tonic firing), sixty-six percent were regular or irregular bursters, and the remainder had no regular behavior and could not be categorized.

Because she wanted to analyze phase response curves, Prinz needed model neurons that mimicked the behavior of a known biological neuron. She chose the PD neuron and scanned the entire database, progressively tightening her search by using successive criteria for the PD's characteristic pattern of bursting. Eighty candidates emerged. Thirty-five of these conformed to the phase response curves she had obtained experimentally. Finally, inspection of the voltage traces of those thirty-five neurons showed that most of them did not look like traces from the biological PD neuron— leaving a grand total of nine model PD neurons for her to work on.

Prinz was pleased with the analysis of phase-response curves her model facilitated. However, in finding her PD neurons, she had demonstrated that the database could be used to identify other model neurons with the characteristics of any specific biological neurons. Marder instantly recognized it as a welcome escape from the tedious business of getting models to work by trial-and-error searches for useful parameters, such as maximal conductance values. She turned the report into a methodology paper and titled it "An Alternative to Hand-Tuning Conductance-Based Models." Prinz says, "I wasn't opposed to it, but I thought the title was almost selling it short because that's only one application of databases. But Eve thought and probably she was right, that it would make sense to a lot of people—a powerful thing that you can do with databases to avoid all that cumbersome hand-tuning."

It was a pioneering approach, and the database was made freely available from the lab's website to other researchers—on sets of DVDs: "Because of the size of the database, downloading it from a website is impractical." It is droll, in hindsight, to note that the model produced fewer than 30 GB of data, which I've got on my smartphone today. At the time, Marder and Prinz were relieved to see that valid results could be obtained from a big database with the computational power available to the lab. The model had to be efficient in using capacity; it was a single compartment model with a choice of only six values for each conductance computed. This sample is "sparse" because it does not include any points in between the six chosen conductance values, unlike the unbroken range of values possible for each conductance in the real neuron. If a "denser" sample had been used around the values that best represented the PD neuron, there would have been many more than nine models. However, even represented by rather coarse sampling, the results were orderly and without discontinuities. This finding was reassuring because Marder knew that in the natural world, any discontinuities would require neurons to avoid the possibility of a dramatic change of activity by some difficult to imagine mechanism of vigilant monitoring and regulation. Instead, the model reflected criteria relatively tolerant of variation but good enough for the animal's requirements.

The database could be screened to find model neurons behaving in a similar way to biological neurons or searched for specific combinations of the neuron properties it contained, such as activity type, phase-response curve, or resting potential. The model could be altered, improved, extended, or progressed to a multicompartment version; the database aspect was a valuable advance.

* * *

Having introduced brute force to their computational work, Marder and Prinz now embarked on another ambitious project: modeling a network. The lab's previous models looked at only one or two neurons together. Prinz wanted to investigate the pyloric rhythm, which, being triphasic, necessitated modeling at least three neurons and representing the synapses between them—as well as the membrane properties of each neuron, a model of a circuit would obviously have to take its connections into account.

Prinz simplified the pyloric circuit in ways that the lab's work had shown were not too much of a distortion; the model could still be judged semi-realistic. The pacemaker kernel, which consists of one anterior burster (AB) and two pyloric dilator neurons (PDs), was reduced to a single burster neuron (but the model did represent both the fast glutamate and slow acetylcholine inhibitory synapses on follower neurons that Marder and Judith Eisen had labored to distinguish twenty years earlier). The several PY neurons were collapsed into one. The lone LP neuron stood for itself. The ventricular dilator and inferior cardiac neurons were omitted. That left a three-neuron network that retained the seven chemical synapses of the pyloric circuit and could produce the characteristic LP-PY-PD, LP-PY-PD rhythm as the lateral pyloric, pyloric and pyloric dilator neurons sequentially release each other from inhibition (see figure 9.1).

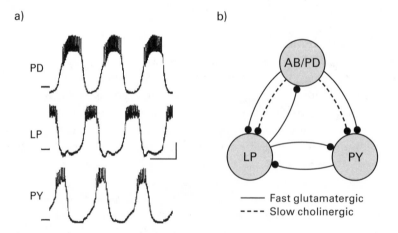

Figure 9.1

(a) The pyloric rhythm recorded intracellularly in *Homarus americanus*, showing the triphasic bursting sequence LP-PY-PD. Horizontal level lines indicate −60 mV. Scale bar indicates 1 second and 10 mV. (b) The reduced version of the pyloric circuit. Synapses are all inhibitory. (After Prinz et al., 2004.)

A moment's patience with numbers is called for here. From her database of nearly 1.7 million neurons, Prinz took five versions each for the AB/PD and the LP and six for the PY (a version meaning one of the sets of conductances that produced activity similar to the biological type). This astute precaution prevented the model from showing idiosyncratic behavior peculiar to a specific single choice from the database to represent each neuron. Thus, there were 150* combinations of the different versions of the three neurons. The seven synapses could be independently varied, and each was given a range of five or six strengths. Taking a deep breath, all combinations of the synapses and their strengths for all combinations of the model neurons resulted in 20,250,000 model circuits.

Naturally, Prinz used her previous model as a basis so that she wouldn't have to hand-tune anything; all the same, it was a brave enterprise. Prinz searched this database for circuits that could generate the triphasic pyloric rhythm. She found 4,047,375 of them, about twenty percent of the total. But the triphasic rhythm was just the minimum requirement. As in her previous work, Prinz had to narrow the criteria in the search; for that, she needed to use the characteristics of biological models of the pyloric rhythm. The experimental data were provided by her fellow postdoc, Dirk Bucher, who made electrophysiological recordings of pyloric rhythms from ninety-nine lobster preparations. Many characteristic features could be identified in his results, and they selected fifteen important ones. Prinz then searched for model circuits that reached the triphasic rhythm, with all fifteen important values falling inside the range found experimentally. Now there were 452,516 "pyloric circuits," about 2.2 percent of the total.

These "realistic" circuits emerged from disparate cellular and synaptic properties. Talking through the results together, Marder asked Prinz to show her two circuits from the database that produced a similar rhythm but where the membrane and synaptic conductances for each neuron varied at least threefold between the two. She was stunned by how quickly Prinz found them. "I took that to mean there were lots of examples; she chose one particular example to make that figure, the one that we always use, but there'd be many, many other examples. Because she found it too fast, right? There were 450,000 valid models in the data base and so, if it had been hard to find, it would have taken her a lot longer." It was a triumphant moment of validation similar to her recognition of the importance of the figure in Goldman's work four years earlier that showed the same behavior in two model neurons with different conductances. This time, it was

---

*$5 \times 5 \times 6 = 150$.

the same behavior in two model *circuits* of neurons in which the conductance sets were different, indicating multiple solutions at the network level (see figure 9.2).

Marder realized that Prinz's work was revealing exactly what she had predicted. In fact, all of the 150 possible groupings of model neurons, using some blend of their intrinsic and synaptic properties, were represented among the 452,516 "pyloric" circuits. It was a dazzling result and bears repeating the other way round: no combination of the different versions of the three neurons was unable to produce a pyloric-like rhythm. That so many ways of producing good enough behavior could be found was resounding support for the multiple solutions concept. Marder and Prinz immediately revised the purpose of their project; rather than an examination of the pyloric rhythm, it became an examination of the multiple solutions that could lead to a similar rhythm, explicitly to "determine how tightly neuronal properties and synaptic strengths need to be tuned to produce a given network output."

The paper reporting this work, published in *Nature Neuroscience*, was titled "Similar Network Activity from Disparate Circuit Parameters." Not for nothing is it one of the most cited of Marder's publications. As she says, "That 2004 paper made what we already knew so dramatic." The finding that "virtually indistinguishable network activity can arise from widely disparate sets of underlying mechanisms" was theoretical, but it was based on robust biological data, and so it now became reasonable to suppose that there was considerable leeway in the animal world too and thus that the phenomenon of animal-to-animal variability should be fully recognized.

For once, Marder allowed herself a generalization beyond the stomatogastric ganglion, writing that the conclusion was "relevant not only to the nervous system, but also to biochemical and signaling networks, as parallel and interacting pathways also occur in these networks."

In the paper, Marder and Prinz made four lucid predictions of phenomena that might be revealed in future biological investigations: first, that the variability in network behaviors between different animals might be more constrained than the variability in the properties of individual neurons in the networks; second, that each conductance or each synaptic strength in its neurons need not be narrowly constrained for a network to be viable; third, that patterns of network output would be found to depend on the correlations between underlying values of synaptic strengths and membrane conductances in its neurons, not on any absolute values; and fourth, that the effect of variable properties of a network's neurons would be mitigated by compensatory changes in other properties.

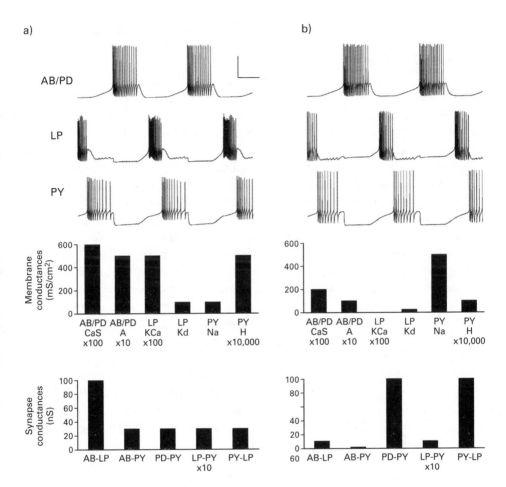

Figure 9.2

Two model circuits, (a) and (b), produce a similar rhythm but each conductance varies at least threefold between the two. The top three traces show the electrical activity of the three "neurons" making up each model pyloric circuit. The upper of the two bar charts shows the membrane conductances for each of the model neurons. The lower bar chart shows the synaptic conductances. (After Prinz et al., 2004.)

In her study of homeostasis, Marder had been thwarted in the search for an experimental demonstration of homeostatic regulation of networks; the recovery of function example had turned out to be unreliable, and no other model had come to light. She had been looking for a parallel at the network level to the homeostatic mechanisms that seemed to operate at the level of the individual neuron; now she thought that perhaps the homeostatic mechanism was regulating network performance itself rather than the intrinsic properties of individual neurons. "For an animal, it is presumably far more important that its behavior be appropriately regulated than that any given intrinsic or synaptic parameter have a specific value."

A network's target performance must be one that produces the essential effect, the behavior essential to the animal's survival. There would be little latitude in maintaining that, which is why Marder and Prinz had surmised that less variability would be found in network characteristics than had been found in the properties of individual neurons. Extrapolating from the models, the results meant that, in the animal, the network was constantly tuned to target performance by neurons that tuned themselves, compensating continuously for each other to achieve a functional balance. It makes sense when you think of each neuron's constant turnover of channels, receptors, and structure, each neuron's cellular processes and responses to its environment; one can imagine that it would be impossible to maintain a perfect state at all times. To Marder, ever alert to the needs of the animal, this looked convincing; after all, it is the animal's behavior, not its individual neurons, that leads to its survival.

The paper ends bravely: "The challenge for future work will be to uncover not only the structure of the networks, but how a target level of network performance is encoded and maintained." Marder, of course, was already taking up this challenge.

At the outset, she asked the question of what characteristics of a network constitute its target—what it is about a network's function that is indispensable to the animal. In the case of the stomatogastric ganglion's networks, what characteristics are fundamental to producing appropriate muscle movements, appropriate behavior?

One way of finding out, Marder thought, would be to examine the development of network function in very young animals because if a value was consistently found in both the young and the adult, that might be a clue to its importance. Similarly, if she could distinguish a network property that was more consistent than others among adult animals, some sort of set point perhaps, that would be a significant discovery. As a corollary, the more variability found in any aspect of a network's performance, the less likely it would be that that aspect was essential.

As it happened, some of the researchers in the Marder lab were just then studying the development of the stomatogastric ganglion in lobsters. It was already known that, even before hatching, the embryonic lobster has the same number of neurons in the stomatogastric ganglion as adults. Marder now speculated that once the juvenile lobster had emerged from the larval form, looked like a miniature adult a couple of centimeters long, and obtained its food in the same way as an adult, it would have to be able to rely on the same mechanisms. Were the output patterns of the stomatogastric ganglion tightly regulated? Would the properties of the participating neurons, although much smaller in size, membrane surface, axon length, and channel and receptor populations, nevertheless be balanced so as to control the workings of the pyloric muscles in the juvenile pretty much as in the adult? Or did the juvenile at first use some approximation of adult mechanisms, or even some different network properties that changed as the lobster grew to maturity?

Marder also wanted to know what the range of variance of network outputs, specifically of the pyloric circuit, between individual adult lobsters might be. Were multiple solutions to be found at network level as well as at neuron level? It was all very well to imagine that network target performance would not have a range equivalent to what had been found in individual neurons, but she wanted to quantify it.

To study these questions, Dirk Bucher and Astrid Prinz examined their ninety-nine adult lobsters and twelve juveniles in detail; the stomatogastric ganglion of each was measured in length, width, and depth, and the surface area and volume were calculated. The morphology of neurons was compared (see figure 9.3).

Intracellular recordings of the principal neurons, the lateral pyloric, pyloric, and pyloric dilators, and extracellular recordings of motor nerves leaving the ganglion were made.

Their rich haul of results included the clues Marder was looking for. The triphasic pyloric rhythm, LP-PY-PD, was found in all preparations, both juvenile and adult. Some circuits ran faster—that is, with shorter cycle periods, or slower, with longer periods, but always with the same phasing: LP-PY-PD. Other attributes of the circuit's output, such as duration of bursts and number of spikes in bursts, were more variable, both between juveniles and adults and between individual preparations in both categories.

Thus, for the animal, maintaining the triphasic rhythm seemed to be more important than its frequency. In that case, could the triphasic rhythm

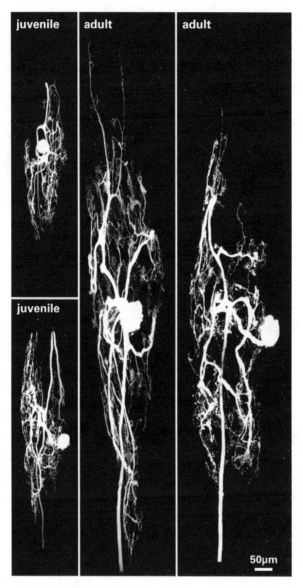

Figure 9.3
Growth of the stomatogastric ganglion. Maximum intensity projections of confocal images obtained from pyloric dilator neurons filled with Alexa 568 hydrazide in two juvenile and two adult stomatogastric ganglia. Note the variability of gross morphological features. All projections are shown at the same scale (see scale bar in bottom right corner). (From Bucher et al., 2005.)

itself be the pyloric circuit's target? How would it be controlled and what feedback would cause adjustments? Clearly, given their earlier data, this consistency of rhythm was not the result of tightly regulated electrical and synaptic values in the individual neurons of the circuit. On balance, Marder and her lab thought it more likely that the *connections* between the neurons of the circuit, with each neuron tuning both its electrical properties and its synaptic strengths, determined the signature rhythm, maintaining the phase relationships. In other words, the controller of the rhythmic pattern might be the wiring diagram.

The work with Prinz was profoundly fulfilling for Marder; it confirmed her intuitions, and she could now make the big intellectual leap to stating that the concept of multiple solutions emerging from underlying variability was indeed credible. The big database of models within which behaviors of interest could be found and examined constituted a complete shift in modeling strategy. Looking back, Marder now says the work was "transformative," a theoretical approach that changed her way of thinking and influenced the following few years' experimental work.

But this seminal work did not go unchallenged. Allen Selverston, for example, took issue with what he saw as an over-reliance on modeling that threatened to divorce itself from experimental results. Marder was already finding ways to address this problem. Thus, over the next few years, some exceptional laboratory results validated the modeling and the wider world of neuroscience absorbed and accepted the gist of Prinz and Marder's work.

It was also another instance of Marder's ability to people her lab with original, clever, and ingenious researchers. Initially, Prinz had not been an obvious candidate for a job in Marder's lab; indeed, she could easily have done something else altogether. She was a graduate student in a physics lab in Germany. Using the principles of neuroscience and electrophysiology, she was building networks with isolated snail neurons in culture. Her PhD advisor sent her to the 1999 Society for Neuroscience meeting. Prinz dutifully went, although she wasn't particularly keen: "It was really unusual for his lab to send a graduate there. We thought of ourselves as higher physicists rather than neuroscientists." But after meeting so many neuroscientists, she came to the uncomfortable conclusion that her work was, if not trivial, at any rate unsatisfying because there was no biological question behind it. She was merely demonstrating that it was possible to make the cells grow in straight lines, make them branch, and create synapses in geometrically simple networks.

While she was in the United States, she thought she would line up a couple of meetings, speculatively, for possible postdoc jobs. She had come across Marder's publications, especially on the dynamic clamp and the first work on homeostasis, and Prinz admired it; she emailed Marder out of the blue asking to visit the lab. Marder was impressed, and in the autumn of 2000, Prinz joined the lab. When she had been there for about a year, Ron Calabrese visited and Prinz discussed her work—and his—with him. Afterward he said to Marder, "Where do you find these people?" She said, "They find me."

* * *

The next unlikely hire to find Marder was Dave Schulz, and he set her off in a new experimental direction. Schulz had earned his PhD researching the behavior of honeybees. He was intrigued by day-to-day variations in their behavior but frustrated by not being able to trace them back to the nervous system; the mechanisms remained opaque. He started thinking that neuromodulation would influence behaviors, "sort of a global way to change the mood of an organism, the likelihood of whether bees would perform one behavior or another. I mean mood is a lousy way to describe that, but you know! And so I was looking at how neuromodulatory chemicals might influence those behaviors."

Schulz became interested in the stomatogastric ganglion as a model that would allow him to get at the cellular level. He had no idea how prominent Eve Marder was in her field when, acting on a chance suggestion, he contacted her. He had no specific project in mind and really no clue, being a behavioral scientist, about what Marder's lab actually did. Certainly electrophysiology was entirely new to him. Schulz admired his new colleagues' expertise: "Jean-Marc Goaillard's electrophysiology skills were unmatched in the lab; he was a terrific biophysicist. But really you could watch any of them at the rig and they knew just how to tweak things. I would look at Dirk Bucher's anatomical fills of neurons and just be astounded at how beautiful they were and what a keen microscopist he was, as well as an experimentalist. I had a very negative opinion of modeling when I arrived, but seeing the work that Eve was doing with Astrid and with Larry Abbott completely turned that around."

Marder left him to his own devices for some months. Finally, he settled on investigating molecular biology aspects of the stomatogastric ganglion, a discipline in which the lab had no track record, and nor had Schulz. Marder encouraged him to run with it anyway.

The stomatogastric ganglion was not generally thought to be a suitable model for molecular biology. But it had proved itself many times over as a good model for electrophysiology, whereas some of the best genetic model systems, such as *Drosophila* or C. *elegans*, have not, until more recently, lent themselves happily to electrophysiology. (In the early 2000s, none of the approaches like optogenetics that allow much more detailed work in these models was available.) Schulz saw the marriage between molecular biology and electrophysiology as crucial to understanding neural circuits. "Either you could take an acknowledged great molecular system and try to fill in the electrophysiology of the circuit, or you could bring molecular biology to a great circuit. And so many people had done beautiful dedicated circuit breaking to describe the stomatogastric ganglion. So it was clearly better to use that circuit and bring in the molecular biology, which these days is mostly in pre-made kits anyway. There's a lot of skill and nuance that goes into it, but it's not like electrophysiology. Electrophysiology is an art form!" This turned out to be the right way of looking at the question; Schulz and Marder were on to a winning formula.

More months passed. Their initial approaches were unsuccessful. Neither Schulz nor Marder was especially surprised, given their lack of experience in molecular biology. Nevertheless, she backed him serenely. "I'd come to her and say, 'I think we should try this but I have no idea if it's going to work and it's expensive, do you really want to do it?' And she would say something like, 'Well, did so-and-so in such-and-such lab get it to work?' And I'd say, 'Yeah, they got it to work.' And she'd say, 'Then you can get it to work. So go do it.'"

Schulz started with ion channels, cloning them from lobster stomatogastric neurons and working out how to quantify them. Probably, he thought, he could use the real-time polymerase chain reaction (qPCR), which was already, in the early 2000s, a mainstream technique. Marder said, "Great. Go get it done."

The first channel that Schulz cloned was the lobster sodium channel; he never did anything much with it, but it got the ball rolling, and that was a relief to him. He had been in the lab for some months with no results, and he wasn't used to that because in behavioral work the observations are immediate. Moreover, he was surrounded in the lab by electrophysiologists getting their data in real time day after day while he sat at the bench moving small amounts of colorless liquids between tubes.

After cloning some channels and genes of interest, Schulz had a nice set to work with. Marder wanted to quantify DNA or RNA expression of ion channel genes in single cells. Soon Schulz and Marder defined the question they wanted to ask: whether the variability in conductance levels (governed by the numbers of each type of ion channel in a cell's membrane) detected by electrophysiology was a reflection of the molecular biology of the animals—of the genetic expression of the proteins making up ion channels. The idea was to discover the range of values for channel gene expression in neurons of the same type in different animals. The lab started with the great advantage of being able to identify, unerringly, the cells of interest (see figure 9.4).

The cells of the stomatogastric ganglion are large; in the crab, for example, the cell somata of LP and PD neurons are about 100 μm in diameter. This size allowed the quantification of several different channel genes in each neuron. With a large surface area—the membrane—the numbers of ion channels in each cell are high. This was another advantage over researchers working with mammalian single cells who had to preamplify their DNA or RNA before they could do anything with it, which could potentially introduce confounds into the quantification. Schulz did not have to use amplification and could better measure expression levels in single cells. He says, "It's still not trivial to be doing molecular biology at the single cell level and that was really Eve's vision of how this should be done. Deb Baro in Ron Harris-Warrick's lab had done some work, not in real time, but their paper was the first single-cell quantitative PCR paper that I had ever seen, and I fell in love and thought it was great."

Using crabs (*Cancer borealis*) and focusing on the LP neurons, Schulz worked with Jean-Marc Goaillard, who made extracellular recordings from the motor nerves leaving the ganglion and measured three potassium ion currents in the neurons. Schulz then extracted them to quantify the mRNA expression of those potassium channels. They found that, for two of the

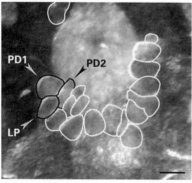

100 μm

Figure 9.4
The stomatogastric ganglion of the crab *Cancer borealis* showing the two pyloric dilator cells and the single lateral pyloric cell. (From Schulz et al., 2006.)

three potassium currents measured in the LPs of nine crabs, the range of the mRNA abundances was similar to the range of values for the measured current—threefold. "I remember the day Jean-Marc and I sat down at the computer and we combined my molecular RNA copy numbers with his electrophysiology measurements of the currents in those exact same cells. We plotted a correlation that showed the current very beautifully tracking the mRNA levels in single cells, and when I saw that plot, I was dumbfounded. I said to Jean-Marc, 'There's no reason that that should work, are you sure we got the numbers right?' It was really too good to be true, but of course, we've repeated it since" (see figure 9.5).

They also plotted the levels of mRNA expression for each channel against the others and found significant correlation between two of them in all cells. To check whether these results were specific to LPs or would be found in other neurons, Schulz and Goaillard examined the same conductances and ion channel mRNA in PD neurons. The range of variability when comparing between preparations was similar to that in the LP cells. However,

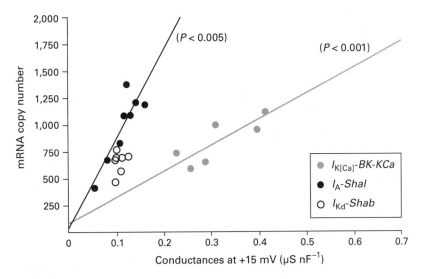

Figure 9.5
In LP neurons, ion channel mRNA abundance (vertical axis) correlated with measured membrane conductance (horizontal axis). Each circle represents a single neuron. Linear regression analysis showed significant correlation between $BK-KC_a$ mRNA and the calcium activated potassium ion conductance $IK(C_a)$ and between *Shal* mRNA and the voltage-gated potassium conductance, I_A. The analysis did not show significant correlation between I_{Kd}, the delayed rectifier potassium conductance, and *Shab* mRNA abundance. (After Schulz et al., 2006.)

they noted with interest that, in the pair of PDs found in each crab, there was less variance and both the conductances and copy numbers for two of the channel mRNAs were similar to each other, although they were significantly different from those in other animals. They speculated that this was caused by some coregulation mechanism between the electrically coupled PDs. Even more interesting, the expression of the channel genes turned out to be correlated in a different way from that in the LP neurons, as you might expect given that the two cell types have different roles in the ganglion with different functional output. What was unexpected was that the two types of neuron were so entirely distinct in their pattern of channel expression—there was no overlap between them, as shown in the three-axis plot in figure 9.6.

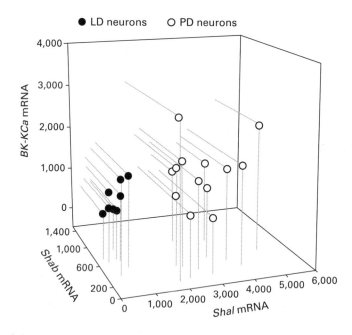

Figure 9.6

Three-dimensional comparison of levels of channel expression in pyloric dilator neurons (PD), open circles, and lateral pyloric neurons (LP), black circles. The mRNA abundance for three potassium ion channels (*Shab*, *BK-KCa*, and *Shal*) did not overlap, revealing distinct expression profiles for the two types of neuron. (After Schulz et al., 2006.)

"Variable Channel Expression in Identified Single and Electrically Coupled Neurons in Different Animals" was published in *Nature Neuroscience* in 2006. The paper explicitly posed a number of questions that had been lurking in the background of the acceptance of variability as a phenomenon. Were the ranges of values they had found in their experimental work indicative of what biology could use, and therefore unremarkable and possibly ubiquitous, or were they instances of something remarkable, a comparatively high level of variability? There were few comparators. What role did compensation play? How significant a factor was genetic variation in the wild crabs they had used? There was comfort in that the results did counter one doubt convincingly: the distinct patterns and correlations in the values, with less variance between pairs of electrically coupled neurons and with different expression profiles in different types of neuron, undermined the criticism that variability was just "noise." The variance observed was neither haphazard nor so great, apparently, as to destabilize the neuron's function, nor so constrained that no leeway was left for tuning to target. Just good enough—allowing neurons and nervous systems to achieve stability while responding to life's challenges. Those crabs had been successful survivors of life in an unforgiving ocean; the variability could be explained by differences in genetic heritage and by different life histories.

Nevertheless, Marder thinks they would not have believed their own results if they had not paired the RNA measurements with the electrophysiology. "How much would you trust measurements of RNA in single neurons? But here we had a very strong correlation between the conductance measurements and the mRNA measurements and that correlation could not have come about by accident. Because if the variance in both data sets was random and just experimental noise, they wouldn't be correlated. And then we had all these very strong correlations between I_A and I_h,* both in the electrical data and in the models, which made it clear that it could be that way. It was really the models that said it *could* be this way."

This study was important because it measured several ion channels in each of several individual neurons whose type had been unequivocally identified; these data were crucial to highlighting the existence of a range of quantities of the different channels within neurons of the same type from different animals.

Marder was eager to extend the scope of the study. Schulz and Goaillard set out to describe the ion channel gene expression profiles of six different

---

*I_A is the transient outward current of potassium ions. I_h is the hyperpolarization-activated inward current.

cell types, measuring six different channels in each cell, again using single-cell quantitative PCR. This study unambiguously confirmed their earlier work. The title of the published report says it all: "Quantitative Expression Profiling of Identified Neurons Reveals Cell-Specific Constraints on Highly Variable Levels of Gene Expression." Again, they found up to fivefold variance in levels of abundance of the same channel between cells of the same type, and again they could plot specific combination profiles of those quantities, plots that did not overlap with the profiles for other neuron types. Five of the six named neuron types had their own signature pattern of channel expression. (See plate 8.)

In the discussion section, Marder and Schulz wrote, "We argue that cell identity is not a static result of gene expression, but rather a continual balance between compensatory changes in gene expression and coordinated gene regulation that ensures robust output."

Added to her earlier concept of tuning to target, in which the intrinsic excitability of a neuron is homeostatically regulated by activity-dependent adjustments, was the new possibility that, whatever the absolute levels for any one of them, it was the correlations of expression in the neuron's channel genes that maintained its functionality. But what dictated those different correlations? Could they be satisfactorily explained as the effects of activity-dependent processes, different in each cell type, some of which affect the cell's pathways for gene transcription factors and hence its levels of expression? Or do the different levels and patterns of ion channel gene expression stem from underlying differences in their complement of transcription factors or microRNAs, determined during the early development of the animal when each cell's "fate" is settled?

These were striking results, especially from a lab that was new to molecular biology. Just as she had taken up other new techniques before, Marder had harnessed molecular biology to her purpose, avoiding the minutiae and simply asking what it could tell her. But Marder shrugs off the success, saying, "I don't think I had any particular expectations of what we would get at the single-cell molecular level because I had no expectations about how long-lived the mRNAs would be or how long-lived the channels would be. I actually had no real expectations. We just wanted to see if we could measure multiple channel genes in the same cell."

* * *

Over the decade, the range of variance confirmed in biological measurements had astounded Marder. The two- to fivefold variability that Jorge Golowasch had found was in data resulting from experimental perturbations. Marder had supposed that in undisturbed conditions, the range might be fairly minor, say up to twenty percent. She was therefore astonished when the evidence in the mRNA studies, without perturbation, also pointed to a two- to fivefold range. With hindsight, she says, "We weren't as upset by the variance in data as other people might have been because we recognized it. So when we saw two- to fivefold differences, we weren't horrified. But we weren't looking for them. We would have taken whatever we got. Because I had no expectation, literally, about how big they would be. If they'd been thirtyfold, I would have shrugged. If they'd been a thousandfold, I'd probably have been very nervous. But they turned out to be, in a sense, in that sweet spot. And that sweet spot is pretty general now, it's what everybody sees. Up to fivefold."

Around that time, Marder met Larry Abbott for a coffee in New York. She asked him how tightly regulated he thought the parameters in a biological network would have to be for good enough behavior, how different the experimental data would be—five percent, ten percent, a hundred? Abbott guessed five percent. "I told him five*fold*. He said, 'No!' He would not have guessed. So it was quite recently that even someone with as much intuition as Larry—he didn't disbelieve me, but he was very surprised." Abbott was of course perfectly familiar with the even wider ranges of variability that could produce 'functional' results in models, but, like Marder, he would have expected the demands of physiology to put greater constraints on the properties of living neurons. The different purposes of life and modeling explain this: real life asks a system to do a lot of things at once, far more than models can yet simulate. Models include only a selection of the biological parameters and are intended to answer a specific question, to do one thing at a time. If the 450,000 "realistic" pyloric circuits in Prinz's model had been asked to do more than merely exist, things would have looked different. If the model had been interrogated to see whether they responded correctly to various perturbations, then there would have been perhaps a few tens of thousands, and even those might have been a more narrowly defined subset or restricted version of the circuit. The more different things a system has to do, the greater the constraints on its path through solution space.

However, the two- to fivefold range is small enough to keep within a reasonable bowl of parameter space but large enough to avoid tuning at a fine level. Any organism would be hard-pressed to tune biological molecules at the two or five percent level; no known biological mechanisms could furnish that level of accuracy and consistency. The two- to fivefold range avoids having to regulate processes too strictly—the transcriptional control of every protein does not have to be perfect—yet is not so lax that compensatory mechanisms are hard to imagine. Biological systems, with their inherent stochastic processes, can stay in that range and produce successful solutions to various perturbations and environmental conditions.

On the subject of animal-to-animal variability, Marder backs away from taking credit for her own insight, saying that the message was also coming from systems biology as researchers in that field started to measure levels of single molecules in *E. coli* and yeasts. "Two communities that didn't talk to each other—and still don't much—came to the same conclusion by very different routes. When I became aware of that work, it took an enormous weight off my shoulders. If you could start with *E coli* or bacteria or yeast, exactly genetically identical, sitting in the same dish right next to each other, and after several cell divisions, or even one cell division, see that the two daughter cells had very different levels of whatever, then how could it be any other way?"

Marder was comfortable generalizing the concept to other systems. "No, I'm not beyond saying that the kind of variability that we see in the crustacean stomatogastric nervous system is going to be a feature of what we see everywhere. And actually now that people feel entitled to report it, of course it's everywhere. Once you convince people to plot all their data, it's everywhere."

* * *

Although surprised by the extent of the variance her lab had measured, Marder was far from unprepared for these discoveries; I have already quoted the passage from her 1974 doctoral thesis in which she considers whether the measurement variability she found might "reflect large actual differences in the activities in the cells" rather than experimental infelicities. Thus, thirty years earlier, Marder had already entertained the possibility that variability would confound her experimental work; now, alas, the glow of achievement came with practical concerns. She knew that experimental practice, in her own and all other biology labs, would have to change. It was the wretched data problem all over again. How you rate your data takes on new importance if you are going to recognize the validity of variability; it is no longer a question of deciding whether to use "best" values or mean values; it is a question of using as much of your data as possible and actually avoiding the use of means as much as you can. This approach went against the grain.

Twentieth-century biology was built on the principle of treating all individuals of a species as though they were identical in order to get meaningful statistics. It was perfectly logical: biologists try to control variables, which is why they have tried to engineer out variability. This paradigm has been convenient because it has allowed confident description of physiological mechanisms and their underlying properties. But the data thus collected over decades have within them a wide range of variations for any one property or process. This variability has mostly been thought of as experimentally caused slack, and experimenters hoped to get close to a true value by statistical methods. Measurements that fell outside a reassuring cluster of values were called outliers, put down to experimental error, and ignored in data that otherwise generally converged giving apparently reliable statistical values.

In her lab, Marder puts the principle into practice: "I don't let my people throw anything away unless they know exactly what went wrong. In other words, if they know that they spat in the dish or the electrode slipped and drove a hole through the cell, fine, throw it away. But if you don't know that you have actually made a mistake, so you don't know why that measurement is an outlier, then you live with it. If you do an experiment twelve times and you messed up once, then you just have to live with it. It's not going to ruin your statistics. If it's a real outlier, then it should be there. If it's an error and you keep it, you're biasing things away from significance, and that's more prudent."

With, say, twenty recordings, throwing away, say, four outliers makes for satisfyingly small error bars; but if the four outliers are kept, then although in many cases it wouldn't change the mean or statistical significance of the result, it will certainly change the variance. That brings a different way of thinking. Marder puts any reluctance to face the issue down to anxiety, to honest concern about getting procedures right, getting quality recordings and measurements.

If a wide range of measurements had to be accepted as valid, if what would have been ignored as an outlying value had to be included, then greater statistical power would be necessary to reveal significant results. The most obvious effect on experimental design would be the need to work on greater numbers of samples. Psychologists have always been aware of sensitivity and power in their experimental approach, deciding whether they expect a big or a small effect from an experimental intervention and planning accordingly. However, Marder says, "Experimental biologists haven't been trained that way, so they don't say do I expect my effect to be big or small, they say how much do I need for significance? So they chase the P-value. And chasing the P is not exactly a very good thing to be doing. In fact it's a very bad thing."

* * *

Recognizing variability and its impact on laboratory practices—the inconvenience and expense of greater sample sizes—was one thing. Of more
consequence was the realization that experimental work was about to get
a great deal more arduous. Plotting the correlations in Schulz's molecular data made it blindingly obvious that finding correlations depended on
making more than one measurement in the same cell. Marder says it forced
her to think about entirely different kinds of experimental design. "If I
could have snapped my fingers and had the complete annotated genomes
of lobsters and crabs, I'd have done a micro array or a chip, a PCR method
that would let me look at two hundred genes, or RNAseq, in a single neuron years ago. And you'd do that on a neuron that you'd already recorded
from electrophysiologically, from a ganglion that you'd already studied, so
you'd be able to correlate the motor pattern to the behavior of the cell to
its molecular underpinnings. That's what Dave Shultz is doing now. In the
best of all possible worlds you want to get as much information as you can
from the same individuals. So you start thinking about bigger integrated
experimental design. And you have to be capable of pulling it off."

What would this mean in Eve Marder's lab? Instead of dissecting out ten
ganglia and testing, say, all the PD neurons for one property, then dissecting out another ten ganglia and testing all its PD neurons for another property, then using statistics and linear regression to make assumptions from
the pooled data, her team would have to find techniques to measure two or
more properties in the same neuron in the same preparation. Researchers
would have to go to greater lengths to make multiple measurements in the
same preparation, trying to get as much data from each preparation, each
cell, as possible—a fuller description of each element of the network. It
would be demanding but would provide extremely powerful results, revealing the interrelationships between different properties in each cell.

Marder's summing up of the new experimental landscape was that she
would be looking for "a whole series of very complex multidimensional
correlations and compensations. It's not that parameter z can be anything
and that parameter x can be anything, it's that parameter z and parameter
x may need to have a certain relationship to parameter a. And so we'll have
to determine for a whole lot of the important parameters—or what we consider important—what the ranges are and how those values are controlled
with regard to all the other things that are important. We have to start
thinking in multidimensional space. Which means it's not good enough
to measure—the sad thing is—it's not good enough to measure synaptic

strength in twenty preparations and measure the excitability of the nerves in twenty more preparations, you have to get synaptic strength and excitability in the same preparations. Because what a synapse will do to the cell depends on how the cell behaves as well as what the synapse is doing."

She was looking at a bleak prospect of very difficult work. Of course, recently the difficulty has come to look less forbidding as new molecular and optical tools seem to appear, one after the other, as if by magic. But this was 2006. In the immediate aftermath of the work with Schulz, Marder and her team took up the challenge with the methods available to them. I have picked two papers as representing the low and the high points in 2007 to 2008, one demonstrating the problem of experimental design and the other exploring the consequences of variability on the stability of responses to neuromodulation.

The Marder lab attacked the issue of multimeasurement experimental design, going on a demanding expedition from which they came back with all sorts of data and a sense of foreboding. In each of sixty-nine preparations of the crab stomatogastric ganglion, they measured and analyzed the timing of the pyloric circuit: the periodicity of its cycle, the onset and offset of bursts, the duration of bursts, and the interspike interval of LP neuron bursts. They measured cellular and synaptic properties of neurons in the ganglion, focusing on the pacemaker AB and PD neurons and on the LP. They recorded synaptic currents in detail, and they looked for correlations between mRNA levels of channel gene expression and the timing and duration of the LP neuron's activity. In short, they did an impressive amount of hard experimental work and a great deal of meticulous analysis of the data. But the report published in 2009 falls flat; the point of it all had somehow eluded even Marder's gifts for making sense of data. "Functional Consequences of Animal-to-Animal Variation in Circuit Parameters" doesn't quite live up to its name. The variability of the parameters measured is described and the correlations found are discussed, but it falls short of the stated goal of understanding how the variability of multiple circuit parameters might be related to the variability in circuit output. No strong story emerges, and the discussion comes closer to waffle than any other paper from the lab I have read. The paper ends, "the challenge is to discover new experimental strategies, such as the multidimensional measurements used here, that can assess which neuronal and circuit elements are variable but functionally critical, and which are only loosely controlled, as they are less essential for the animal's behavior." Marder referred to it as "despairing," an apt description.

The second study that stands out in those years addressed the question of whether neurons of the same type but with different intrinsic properties nevertheless respond to neuromodulation in the same way. What were the consequences of the significant range of variance to be found in each neuron for its responses to perturbation or neuromodulatory stimuli?

The Marder lab had already researched the analagous problem for single neurons. Mark Goldman had established a theoretical framework for the different effects of neuromodulation on neurons with different intrinsic values. Turning to the network level, the question became: if a circuit in one animal consisted of neurons that differed in their intrinsic and synaptic states from those making up the circuit in another animal of the same species, would the two circuits respond to neuromodulation in the same way? With a new graduate student, Marder found out that, on the whole, they would. In each of her preparations, Rachel Grashow set up a "circuit" consisting of two of the four gastric mill (GM) neurons in a crab's stomatogastric ganglion. The cells were isolated from any input other than an artificial membrane conductance supplied by dynamic clamp. The dynamic clamp also controlled the strength of the reciprocal inhibitory synapses between the two cells. In other ways, each cell's behavior was the product of its intrinsic properties. It was a clever setup, taking advantage of the ability to compute and control two variables while also manifesting the inherent animal-to-animal and cell-to-cell variability. Depending on the settings of the dynamic clamp, the two cells were sometimes silent, sometimes one or the other produced spikes (a behavior they called asymmetrical), or both cells spiked, or both produced bursting in alternation. Each two-cell circuit was tested with an application of serotonin and later with oxotremorine, two neuromodulators known to activate different conductances in GM neurons.

The results of the manipulations and the behavior of the mini-circuits in saline, serotonin, and then oxotremorine were mapped. These maps were revealing, showing that in conditions where alternating bursting had been seen, both of the neuromodulators increased the frequency of the bursts, with the notable exception of a few instances that were contrary and tended to slow down (see figure 9.7).

The finding that circuits with different underlying parameters responded canonically to neuromodulation also revealed that what had been taken as well-known and expected responses to these two neuromodulators in fact concealed individual inconsistencies. In the 2009 paper describing this work, Marder called the nonconforming mini-circuits "anomalous." For this date, it is an unusual research paper, as distinct from a review, in

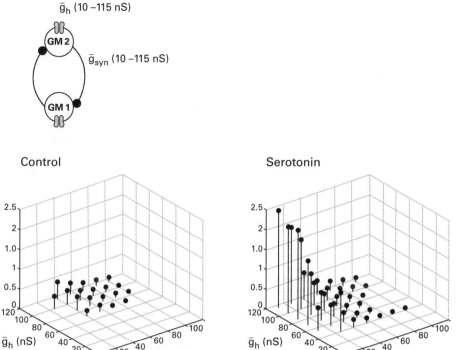

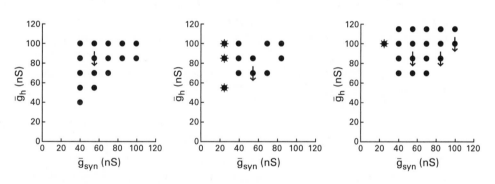

Figure 9.7

Top: Diagram of the circuit of two gastric mill (GM) cells in which a dynamic clamp could control the membrane conductance, g_h, and the strength of the inhibitory synapses, g_{syn}. Center: Three-dimensional plots of burst frequency against membrane conductance and synaptic conductance for those circuits with half-center activity show that the presence of serotonin increased both frequency and the range of g_h and g_{syn} values that produced bursting activity. Bottom row: Data from three of the two-cell circuits with reduced bursting frequency in serotonin. The combinations of parameters that produced this anomaly, marked with down arrows, are different in each case. In some conditions, these circuits produced single-spike bursting, shown here by asterisks. (From Grashow et al., 2009.)

that it was written by Marder herself. Its bold style contrasts markedly with the pessimism of that other key publication in the same year. In her conclusion, Marder firmly puts the work in the context of human brains and nervous systems: "These data provide insight into why pharmacological treatments that work in most individuals can generate anomalous actions in a few individuals. ... Some individuals will respond anomalously to drugs, food, or sensory stimuli, even when their baseline behavior may be within the normal range. Untoward responses to pharmacological agents may occur when individuals contain circuits whose underlying parameters combine to make them particularly susceptible to certain kinds of perturbations, although each of the underlying parameters could be within normal ranges."

In Marder's thinking about homeostasis, multiple solutions, and variability, the importance of mechanisms of compensation had come to dominate. Cracking the codes of compensation would be the big story that would show how to link concepts of variability and individuality to the rules of physiology, how to understand the boundaries between healthy physiology and disease, and therefore lead ultimately to personalized medicine.

It was difficult to find experimental ways of seeing into the workings of a biological network. With Rachel Grashow, Marder devised a hybrid two-cell circuit, one cell being a real neuron in the stomatogastric ganglion of a crab and the other a model neuron. Using the dynamic clamp, membrane and synaptic conductances could be varied—and of course recorded. So the resulting behavior of the two cells would take place in a transparent network.

In each stomatogastric ganglion, Grashow measured six of the electrical properties* of four cells of different types: lateral pyloric (LP), gastric mill (GM), dorsal gastric (DG), and pyloric dilator (PD). The properties for each cell type were variable as expected. Then, still keeping each biological cell in place in the ganglion (and blocking any input from it), the cell was connected by dynamic clamp to a model of an oscillator cell. This made a neat little reciprocal inhibitory network in which the dynamic clamp controlled the biological neuron's membrane conductance and made an artificial reciprocal synapse between it and the model cell (see figure 9.8).

It showed that there were settings of the dynamic clamp that would certainly not have produced useful circuit behavior, but some settings resulted in the two cells firing in alternate bursts, called half-center oscillations,

---

*Input resistance, spike threshold voltage, spike height, frequency-current curve, spike frequency, and minimum voltage.

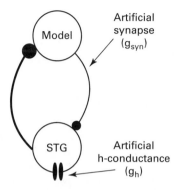

Figure 9.8
Diagram of the "network" linking a model cell with a biological neuron in the stomatogastric ganglion. The simulated inhibitory synapses and the simulated hyperpolarization-activated inward current conductance I_h were supplied by the dynamic
clamp. (From Grashow et al., 2009.)

reflecting real-life circuit dynamics. This was true in each of the twelve LP
cells tested. Grashow made three-dimensional plots of her results, and these
showed a distinct region of parameter space, a clustered group of conductance values and synaptic strengths—in a two- to threefold range—that
produced this activity, whereas other values did not (see figure 9.9).

Using her measurements for the LP cells and taking as target the pattern
of alternate bursting within a realistic frequency range, Grashow found the
dynamic clamp input that produced the pattern. For each LP cell, with its
particular starting electrical properties, a particular combination of membrane and synaptic conductances provided by the dynamic clamp was
necessary. The ensuing plot is illuminating. (See plate 9.)

Each LP neuron had reached the common target activity pattern, but
each did so with different dynamic clamp settings. This suggested that compensations for the variability in each neuron might allow it to achieve a
target activity pattern. Thus, the essential network activity could be reached
with variable compensation for the variable intrinsic properties between
the cells.

For years, Marder's prediction had been that the concept of homeostasis
implied mechanisms of compensation, that homeostatic processes could
find multiple solutions to producing good enough behaviors in both single
neurons and networks, provided there were also processes of compensation. Describing and interpreting those processes would be the next step.

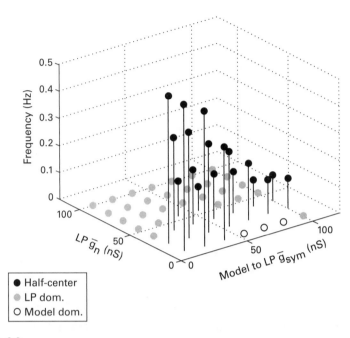

Figure 9.9
For a biological LP neuron connected as in figure 9.8, a three-dimensional parameter space map shows a clear demarcation of the synaptic and conductance values that produced half-center activity. Black data points denote networks that achieved half-center activity; grey points denote networks dominated by the real lateral pyloric neuron; at circled points, the network was dominated by the model neuron. (From Grashow et al., 2010.)

10 In the Big Picture

In September 2008, Marder was a guest speaker at the Inaugural Kavli Prize Ceremony and Symposia in Oslo. Arthur Wingfield went with her. "Eve gave a talk about Astrid's zillion simulations of which thousands produced the triphasic motor pattern. Somebody in the audience said, 'Well, you say there are all these possibilities, but are they all equally good?' And Eve paused for a nanosecond. 'Not necessarily, but there might be some perturbations—say, climate change and the seas changing temperature—when some would actually still work and others not.'"

Ever since Marder had recognized the validity of multiple solutions, she had been mulling over the notion that all solutions might not be equally good. As soon as Astrid Prinz's second paper, "Similar Network Activity from Disparate Circuit Parameters," was published in 2004, Marder found herself fielding reactions from colleagues. If there are so many possible solutions, then how can the animal always find a stable, adequate response to perturbation? The idea ran counter to the pervasive belief that each species had arrived at an optimal physiology and that its phenotypes were as close as possible to this optimum, a belief founded on the early interpretations of the theory of evolution and on nineteenth-century respect for nature's wisdom. Perhaps every detail of an animal's physiology was just good enough to do its job. In that case, the precise values of, say, a neuron's membrane conductance wouldn't matter as long as there was a good enough outcome at every level to make the next level work. But if optimality was to be replaced by adequacy, it followed that some solutions might be only just good enough, able to function in normal conditions but less resilient to challenge and change. This distinction might be an important clue to the difference between robust health and susceptibility to disease.

In Oslo, on the spot, Marder made an explicit decision to use temperature perturbations to probe that aspect of the consequences of multiple solutions. Temperature, she thought, would be the perfect perturbation to study because it affects all physiological processes. Ironically, it was precisely for that reason that she had avoided working on the influence of temperature thus far. "People always asked me why I didn't study the effects of temperature, and I always said, 'Eeugh! Who would want to? Temperature affects everything; what are you going to learn from it?'"

Back at Brandeis, Marder had a graduate student who was interested in the mechanisms of homeostatic regulation. Lamont Tang went to work, looking at the pyloric rhythm in *Cancer borealis* preparations in temperature controlled saline. Crabs are poikilothermic; they do not regulate internal body temperature physiologically but can avoid extremes behaviorally—by moving to more favorable places or depths. They thrive, in the case of *C. borealis*, in water from 3°C to well over 20°C, sometimes in the same 24 hours. The pyloric rhythm is always active, and so it would be expected to accommodate change within that range. As Tang increased the temperature of the saline, the frequency of the rhythm increased, as anticipated. However, unexpectedly, the phase relationships remained stable (see figure 10.1).

Even more unexpectedly, when Tang overlaid the waveforms recorded from PD, LP, or PY neurons on each other, after scaling them to a uniform cycle period, they were near identical at all temperatures. Marder remembers, "I really was shocked. It was Lamont's idea; he just came into my office with a picture and I said something about holy smoke! He was careful, meticulous, a good physiologist and a clear thinker so he knew what to do, it was very convincing" (see figure 10.2).

Tang and his colleague, Jon Caplan, went on to investigate these phenomena in a computational model. It indicated that an LP-like model neuron could compensate for temperature increase when its intrinsic properties had similar Q_{10} values (see Glossary). Processes that depend on changes in protein conformation typically have higher Q_{10}s; ion channels can have Q_{10}s from 2 up to about 100. Thus, in a neuron, the effects of temperature change are complex and certainly affect excitability. Caplan's computational study made it clear that the Q_{10}s of the structures and processes that together produce the pyloric rhythm must all be similar, even within the multiple solutions possible. Q_{10} is therefore a parameter that restricts the range of possible solutions for good enough behavior under conditions of rising temperatures.

To find out whether a crab's life history affected its responses to a range of temperatures, three groups of crabs were first acclimated over at least three weeks at three temperatures within the range of the crab's habitat:

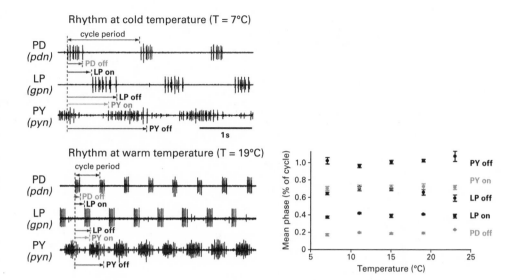

Figure 10.1

Effect of temperature on the pyloric rhythm. Upper panel: Extracellular recordings from nerves showing the frequency of bursts from pyloric dilator, lateral pyloric, and pyloric neurons in saline at 7°C. Scale bar: 1 second. Lower left panel: From the same preparation, showing higher frequency at higher temperature (same time scale). Lower right panel: The phase of the pyloric rhythm is unaffected by temperature. Data from seven preparations showing the mean phase (delay divided by the cycle period, as defined in lower left panel, at each temperature). (From Tang, et al., 2010.)

cold, 7°C; medium, 11°C; and warm, 19°C. Because Tang's first study had shown that the frequency of the pyloric rhythm increased with temperature, it seemed likely that the temperature to frequency relationship might be affected by acclimation. But acclimation made no difference at these temperatures, and this result supported work by Dirk Bucher and Astrid Prinz from a few years earlier, suggesting that the target for homeostatic regulation is the triphasic pattern, not frequency.

Looking for a way to reveal hidden vulnerability in some phenotypes, Marder and Tang turned to the challenge of acute high temperatures. In experiments taking the temperature of the saline up to 23°C—still a temperature that the animals might encounter—all the preparations responded similarly regardless of acclimation. The frequency of the pyloric rhythm increased about fourfold over the temperature range, and phase relationships were maintained. However, the preparations from cold acclimated

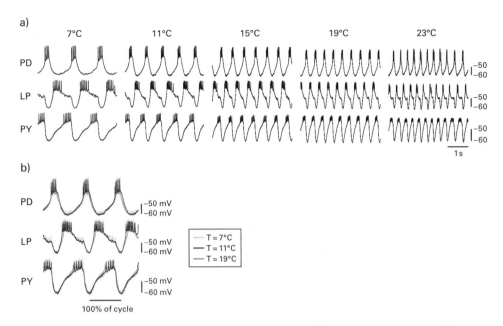

Figure 10.2
Upper panel: At the same timescale, simultaneous intracellular recordings from pyloric dilator, lateral pyloric and pyloric neurons show similar waveforms at different
temperatures. Vertical scale bars, 10 mV. Horizontal scale bar, 1 second. Lower panel:
Traces recorded intracellularly from PD, LP, and PY neurons in a single preparation
at 7°C (light gray), 11°C (black) and 19°C (dark gray) scaled for cycle period and then
superimposed upon one another. Vertical scale bars, 10 mV. Horizontal scale bar,
1 duty cycle. (From Tang, et al., 2010.)

animals started to show stress at 23°C, with less consistent rhythm, whereas
warm acclimation seemed to provide greater resilience at acute high temperature, extending the range for good enough behavior. But when the
temperature of the saline was increased to 27°C and 31°C, even the warm
acclimated preparations were stressed. At these temperatures, which a North
Atlantic crab is most unlikely ever to encounter in the wild, the pyloric
rhythm was disturbed, and more animal-to-animal variability was seen. At
31°C, most preparations dropped out, losing the rhythm completely. Tang
called this state a "crash," joking that it is not a "cook" as the rhythm
resumes when the temperature is brought back down to tolerable levels.
Interestingly, each of the preparations crashed in a different way, with
different dynamics, reflecting the variability of their neuronal properties.
(See plate 10.)

Marder had her answer to the post-lecture questions: although there are multiple solutions to maintaining functions, when animals are exposed to challenging conditions, differences in cellular properties that were of no consequence under ordinary conditions could cause different outcomes. Within the different sets of parameters in a population, some sets might be good enough only until stressed: good enough in normal conditions— those of the animal's habitat—but failing in harsher conditions. Other sets might be better able to respond to extremes. If the conditions of a species' habitat change, then some of these combinations of properties may permit certain animals to survive, allowing a species to adapt to the new environment.

These results added to an emerging picture of the compensatory phenomena that make such resilience possible. Marder took up the challenge of understanding how a system can be at once stable, robust, and plastic. Talking about the lab's work on temperature and its subsequent investigations of performance under changing conditions of pH, she said, "The animals are far, far better at handling environmental perturbations and what goes on in their lives than you ever would imagine feasible given how many different biological processes are involved and how much animal to animal variability there is."

* * *

Over the next few years, various pieces of the puzzles Marder had set for herself were put in place, and some of the questions she had been asking were clarified, if not definitively answered. For example, the long-running and troublesome doubt about whether a recovery phenomenon really existed was reexamined by Al Hamood. A modeling study by Gabrielle Gutierrez highlighted the role of degeneracy in ensuring robustness and resulted in a spectacular way of analyzing nonlinear data sets. Marder successfully revisited the modeling of homeostasis with Tim O'Leary, and Adriane Otopalik applied new research techniques to find detailed evidence that neuronal morphology is not optimized, just good enough. These are only a few of the standout investigations of recent years, and alas the only ones there is space to describe here.

Hamood's meticulous work on the vexed question of recovery of function after the removal of incoming neuromodulatory influences was convincing. With the benefit of continuous saline perfusion during prolonged recordings and of improved data collection and analysis, he was able to show that when neuromodulatory inputs to the stomatogastric ganglion were cut off, the pyloric rhythm slowed within thirty minutes but did not actually cease. Over the several days of recording, this slower activity was maintained more or less steadily, but the ganglia did not recover their previous cycle frequency. Marder now thinks it is not fair to others who have worked on the phenomenon to say that there was no recovery; she prefers to say that Hamood did not see apparent recovery because none of the preparations he tested went silent after being isolated. Curiously, in the summer of 2017, some of the decentralized preparations again went silent. Marder wonders whether the key to seeing clear recovery will be to study only the population of animals that stops or becomes very slow after decentralization. There is no apparent clue as yet to why there are groups of animals that slow but don't stop and others that do stop but Marder is now trying to keep detailed lab records of catch season, ocean temperature, and molt cycle for all experiments, looking for correlations. It is a fine distinction, but in previous studies that did show loss of activity followed by recovery, the restored pyloric rhythm is generally slower and less stable, similar to the activity patterns recorded by Hamood. Possibly that level of performance is the target for recovery, imperfect though it is.

Hamood went on to a similar study of the gastric mill rhythm, which, unlike the pyloric rhythm, is not continuously active in the animal and is termed episodic. Only about a third of the preparations continued any activity after modulatory inputs were removed, and such activity slowed down. Unlike the pyloric preparations however, it was also disorganized: phase relationships were not retained, and individual neurons sometimes fired randomly. The gastric rhythm is thought to be generated by descending modulatory inputs, which are activated by sensory neurons. Thus, removing these inputs may be more significant for the gastric circuit than for the pyloric, there being no identified pacemaker among its neurons.

This is something of a sidetrack, but of course the very existence of episodically generated rhythms poses the question of what sort of activity-dependent regulation of conductances and synapses can be involved after long periods of inactivity. Somehow the activity pattern must be conserved for its next activation, but how can the system be prepared for activity when it isn't required all the time? It is possible that activity-dependent regulation of the conductances in gastric neurons may not follow the same imperatives as those governing the continuous pyloric rhythm. Characterizing homeostasis or homeostatic compensations in episodic systems is something Marder recognizes as a challenge that her lab has not yet confronted. The work would be of interest because the gastric rhythm can be interpreted as representing episodic activity in pattern generation, equivalent to walking or swimming. It is therefore possible that clues may be found to mechanisms of recovery or compensation after spinal injury, which can be troubled by unpredictable spontaneous neuronal activity.

* * *

At a seminar meeting at MIT in June 2013, Marder appeared in an eye-catching, intricately patterned tunic, saying, "I'm wearing a visualization tool that my graduate student, Gabrielle, invented as a way of capturing the dynamics of five cells as a function of multiple parameters, a novel way of displaying data."

Gabrielle Gutierrez came to Marder's lab with a first degree in physics. Looking at the connection map of the stomatogastric ganglion with the fresh eyes of a novice, she picked out a symmetrical arrangement of two pairs of reciprocally inhibiting cells, both coupled to the inferior cardiac neuron (IC). The arrangement lay on that hazy boundary between the pyloric and gastric circuits, what one might call a boundary of convenience, the dream of neuronal circuits neatly delineated from each other and easy to study separately having long been dispelled.

With Marder and Tim O'Leary, Gutierrez thought up a way to explore the activity of five model neurons connected in a similar way. Her intention was not to study the workings of the five cells of the stomatogastric ganglion per se but to extract the connectivity motif and see what she could do with it. By increasing and decreasing the strengths of the synapses, both chemical and electrical, as shown in the figure, the allegiance of the hub neuron could be switched back and forth from the faster rhythm of one pair of cells to the slower rhythm of the other. (See plate 11.) Most important, more than one type of operation could make the model hub neuron change its firing frequency; in other words, the hub neuron was affected by changes in more than one pathway. These would be described as degenerate mechanisms (when two or more different components of a system can perform the same function) or parallel pathways, and they are widespread in disparate fields of science but hard to study.

The model thus underlined the importance of knowing the connections, of having a reliable wiring diagram to support studies of circuit dynamics by identifying the possible parallel pathways and degenerate mechanisms. In her public lectures, Marder now warns researchers against assuming that any pathway under study is the only one. Calling the opportunity for analysis this model had created "a luxury," the report on the model warns, "This luxury is not available when studying biological networks in general, so care must be taken to understand the degree to which the system in question exhibits nonlinearity or degeneracy when the behaviors of its components are summed together. These two generic features, non-linearity and degeneracy, while proving to be obstacles to a mechanistic understanding of nervous system function, also explain its flexibility, the richness of its repertoire of behaviors, and, in the case of degeneracy, its robustness."

Stated like this, the model hints at a clear message, but reading that message and communicating it to others would be challenging. They had a mountain of data covering the activities of the five cells in all the combinations of synaptic strengths in the model. The raw data were, of course, in numeric form. Now Gutierrez invented something entirely original; Marder speaks of her creativity in awe-struck admiration. It was a new way of displaying the data that Gutierrez called a "parameterscape." It has simple rules: the state of all five cells for any combination of two parameters can be shown graphically because, using colors to indicate oscillation frequency, each data point has four concentric circles—one for each of the neurons in the half-center pairs and a square for the hub neuron. The frequencies of all five cells can be shown as a function of the strengths of the electrical coupling and the inhibitory synaptic input. At a glance, one can see what is happening for each combination of synaptic strengths. Even more strikingly, the discontinuities stand out so that the whole picture of synaptic influence yields its secrets and surprises. Marder was delighted with this aid to the problem of comprehending more than two or three changing elements at once. (See plate 12.)

As O'Leary points out, "You can say all that stuff in words. Take any neural network, it's got nonlinear components, multiple pathways, you'd expect lots of different parameters to give you similar behavior, plus lots of difficult transitions between the behaviors. But seeing the specifics makes the point quite differently. From that image you can say, if the circuit is in this region, it's pretty much stable. If it has to get to another region, another type of activity, then varying a parameter in this direction may be one way of doing it."

Tim O'Leary knew and admired Marder's work long before he thought of joining her lab, but he had no introduction when he rang her; it was a cold call. As a postdoc at Edinburgh University, he had been working in a crowded field—using hippocampal slice preparations from mice—and hadn't enjoyed the competitive aspect of it. His background had been in mathematics and informatics before turning to physiology and neuroscience. By 2011, O'Leary wanted to move into computational neuroscience, and Marder's approach to modeling appealed to him. He felt that Marder had recognized the limitations of available data; instead of trying to literally model nature, she was looking for what models could reveal in a generic biological sense, raising new issues or showing how, in principle and at a certain level of abstraction, mechanisms might work.

It was a perfect match. O'Leary joined the lab in early 2012, and nine months later he told me, "One of the first things I noticed was she'd hired really nice people. She told me that was most important; it's pointless having someone brilliant who alienates everyone else around them or lies to them.

"And I've learned about how to structure an idea; I think that, whereas doing quantitative biology or neuroscience is a science, choosing a question is probably a bit of an art; it requires experience and a bit of insight, but also a sort of aesthetic approach, some quality to questions that make them attractive.

"The other thing I'm learning is how to reach an audience that matters. Computational neuroscience should actually help people who are non-computational to think about what they do, putting results and data in context and deciding what the important next questions are. Obviously, I need to speak to computational neuroscientists as well, but they're already speaking my language. It's far more important to speak to biologists."

That same day, Marder told me: "Tim is in the process of going back to square one and doing a whole new set of homeostatic models. He started out as a mathematician; he's really good. We're going to get homeostasis right, which it isn't, neither the theory nor the model."

O'Leary had already been thinking about theories of homeostasis and, with Dr. David Wyllie, his supervisor at Edinburgh University, had published an assured and critical review, "Neuronal Homeostasis: Time for a Change?" in the *Journal of Physiology*. He was interested in applying the control theory that stems from cybernetics to regulatory processes in the cell. Cybernetics is prone to arcane vocabulary, but, in plain English, the theory applies to a dynamical system that has activating inputs and a target output. A sensor reports on differences between the detected output and the target, and these error signals feed back to correct the input.

The new models were, however, not at all what Marder had aspired to twenty years earlier. The model that reflected the entire machinery and morphology of the neuron and the networks of neurons was still unachievable and, by this time, potentially as difficult to understand as the biology itself in Marder's eyes. Unmoved by the siren calls of vastly improved computing power and immense increases in the data available, she had realized that biological replication in modeling had to be severely limited.

When O'Leary got down to work, he chose a level of biologically plausible simplification.

Since the remarkable work done by Astrid Prinz and, of course, because of it, the message of variability had been heard. Faced with choosing data in a necessarily haphazard way, there being so many "versions" of the same neuron, and faced with the multiple solutions adopted by neurons and networks, it was logical to look instead for generic structures in modeling. It was also logical to look for models based on the cell's control of its population of ion channels, of its proteins and genes, rather than to match the quantities of channels found experimentally.

O'Leary started by concentrating on the regulation of the membrane properties of a neuron, looking for a homeostatic control mechanism that might govern the correlations between the expression levels of different ion channels, such as those revealed by Dave Schulz and Jean-Marc Goaillard. It would have been easy to conclude that these correlations were important for shaping the intrinsic characteristics of different types of neuron and therefore shaping their behavior. But the lab's previous computational models, such as Adam Taylor's and Astrid Prinz's, were intended to search for sets of conductances that produced a specific pattern of activity, and these sets, although very varied, did not contain the correlations found in experiments. How did these correlations arise?

O'Leary explained his reasoning to me: "The nervous system needs some principled rule for getting its palette of conductances in the right ratios—or approximately so. But there are far too many possible solutions, so it needs some sort of generic self-organizing plan that gets it right most of the time and in a way that's robust." In O'Leary's self-organizing plan, the model neuron monitored intracellular calcium ion concentration, $[Ca^{2+}]$, as a surrogate of its activity. So far, so familiar. His important innovation was to vary the rates of production of each different ion channel.

Using a term borrowed from physics, O'Leary called his first attempt a "toy model," a simplified prototype of what he was planning. The model neuron was a single compartment. It was a "leak model," meaning that the three conductances he modeled were not voltage-gated; the ion channels represented allowed ions to pass irrespective of membrane potential. It had a mechanism that varied the level of each conductance independently, according to the cell's activity as represented by its Ca^{2+} concentration.

The toy model predicted that this sort of independent tuning of each conductance according to a single feedback regulator would result in correlated conductances, the correlation values being a function of the relative production rates of each ion channel. It proved to be a useful pointer to more complex modeling work. O'Leary applied this paradigm to a more detailed model with seven voltage-gated conductances and ended up by confirming its validity with reference to Zhang Liu's model from the late 1990s, with three different Ca^{2+} sensors.

These models confirmed that he was on the right track, indicating that an activity-dependent feedback mechanism, acting on the conductances at a specific rate for each, would always reach a steady state that contained correlated conductances. "Correlations in Ion Channel Expression Emerge from Homeostatic Tuning Rules" was published in 2013. The discussion is rich in issues to ponder, and I hasten to say that it emphatically does not claim that activity-dependent regulation is the only factor in determining levels of channel expression; other cellular processes are expected to contribute too, further constraining the possible solutions. Nevertheless, the results were broadly consistent with the earlier experimental findings.

It was an excellent paper, but in my opinion overshadowed by the next. "Cell Types, Network Homeostasis, and Pathological Compensation from a Biologically Plausible Ion Channel Expression Model" was published in *Neuron* in 2014. It is hard to explain in a few paragraphs, and even harder to do justice to.

The new model, although again a bold abstraction, conformed to known biological principles. O'Leary's new self-organizing plan explicitly followed the central dogma of molecular biology: a transcription factor activated channel genes, channel mRNA was transcribed, channel proteins were synthesized. It was a stripped-down version, of course; the many complexities of transcription and translation were omitted. Notably, the self-organizing plan reflected the key physiological signaling role of calcium ions in activating gene expression, especially in excitable cells like neurons. O'Leary devised a single transcription factor, representing some sort of Ca^{2+}-activated enzyme, whose rate of production was dependent on feedback from errors in meeting the $[Ca^2]$ target, that is, from Ca^{2+} concentrations above or below it.

The model included seven voltage-gated channel types as well as a leak channel. The production of mRNA for each channel depended on the transcription factor, and each channel protein was produced at a unique rate from its mRNA. Thus, essentially, a feedback loop maintained average $[Ca^{2+}]$ by recurrently adjusting the expression rates of channels for the membrane. Following simulated activity representing, for example, a change of input from other neurons, if the effect on $[Ca^{2+}]$ caused an error signal, compensation was ensured. From an initial random set of low conductances, by setting different expression rate constants for each channel type, the model settled at a steady state, with the relative amounts of each of its ion channels characteristic of a different cell type. (One of the model's shortcomings was that it produced pairwise correlations for all ion channels, whereas experimental work had shown linear correlations between only a subset of channels.)

The coupling between the cell's activity, changes in Ca^{2+} concentration, and expression of each channel assumed slow dynamics, reflecting the biology of a cell in which such conductance changes take place on a scale of hours or days. As the published paper pointed out, this is much slower than the changes in the cell's activity, reflecting the influence of conductances on the cell's *average* activity, rather than immediate responses to perturbations. However, perturbing the model in various ways showed that the model neurons did not always compensate their electrical properties. When, for example, a single ion channel was deleted, the homeostatic mechanism sometimes caused a "pathological" loss of function.

In the same study, O'Leary applied the model to a representation of the canonical pyloric network with three model neurons representing AB-PD, PY, and LP and showed that network-level homeostasis arose in this minicentral pattern generator. Indeed, as O'Leary remarked, "It looks as though all the neurons are coordinating themselves with respect to their neighbors, but actually they're just monitoring their own activity."

Conceptually, this work was a tour de force. Without the support— or distraction—of experimental data, it showed that a simple regulatory pathway representing the principal features of ion channel expression can produce a fair model of activity-dependent conductance regulation. The extreme simplification had the added advantage that none of the premises of the model was species specific. O'Leary airily played it down: "It's the simplest model you can come up with—it is actually DNA to RNA to protein, and then there's calcium dependence. If you just assume that and you assume different cell types have different relationships with those steps, then you're done! We know that the cells are more complex than that, but if we can get that far with such a simple model, then it's not too difficult to understand how, in principle, the cells can control their properties. If the dumbest model in the world can do it, then one that has evolved over billions of years certainly can."

After this, O'Leary and Marder wrote a thoughtful paper in favor of models only loosely founded on experimental data, arguing that such models were useful in revealing fruitful avenues of thought and in pointing to the possible existence of mechanisms for further investigation. O'Leary left for Cambridge University in December 2015. Their collaboration, combining profound and creative mathematical insight with outstanding biological intuition, continues. Marder told me, "He was an extraordinary intellectual force in the lab. I miss him every day."

* * *

The most casual inspection of any neuron under the microscope reveals the challenge of regulating intrinsic membrane properties when all types of signaling can occur anywhere in the branches of its dendritic tree. How distal events affect the cell as a whole is a fascinating question and points to homeostatic mechanisms at local dendritic levels. Within the typical morphology of a given type of neuron, much variability is seen in different preparations. Yet these neurons are fulfilling the same functions. Thus, another of the puzzles that animal-to-animal variability raises is the relationship between individual morphology and appropriate physiology within a cell type.

Over the years, one of Marder's unfulfilled goals had been to investigate the extent of compartmentalization and local influences in different parts of a neuron; it was among the questions she had listed in her lab books during her postgraduate days. In the early 1990s, in the lab Marder shared with Larry Abbott, theoretical work with Abbott's student, Micah Siegel, had served to whet her interest. His model neurons, with the characteristics of a mammalian neuron, had developed non-uniform distribution of voltage-gated conductances in different compartments. (See plate 7.) But the detailed experimental investigation and analysis of neuronal structure and function, taking measurements and monitoring events in the fine neuronal processes in a neuropil like that of the stomatogastric ganglion, was unattainable at the time. More recently, the development of sophisticated experimental techniques has made these levels of enquiry accessible. In autumn 2012, Marder went to a talk by Bernardo Sabatini about work using photoactivatable neuropeptides. It caught her attention, and then, as usual, she waited calmly for the right opportunity, the right researcher.

"Graduate students never get anything done in their second year," remarks Marder drily. Adriane Otopalik arrived in Marder's lab in mid-2012, her second year of graduate school, and conformed to tradition. But by early 2014, she was ready for a challenge and decided to use these photo-uncaging techniques to probe the effect of neuromodulators on different parts of the dendritic tree.

Determined and resourceful, Otopalik rustled up the advice she needed and set about acquiring and assembling the equipment that would enable her to see where receptors and ion channels were clustered on a neuron, where the responses to transmitters and modulators were located. By comparison with the custom-built fluorescence microscope mounted on a manipulator, the UV laser and dichroic beam splitters, the confocal microscope, and not forgetting the software suites, Marder's work three decades earlier trying to track electrodes through the neuropil had been like fumbling in the dark wearing fur gloves.

One of Bernardo Sabatini's postdocs at Harvard Medical School, Matthew Banghart, helped with the setup and the design of caged peptides. These were a disappointment and Otopalik turned to uncaging the neurotransmitter glutamate. The target molecules are "caged" by a chemical modification so that they do not react until a beam of light releases them. Treated molecules in the bath can be manipulated using light; a focused beam can release them rapidly and with precise timing close to just one part of a cell. Otopalik set out to quantify the variability between preparations in the dimensions of processes and the branching patterns of the gastric mill (GM) neuron in *Cancer borealis*. Her goal was to discover how animal-to-animal variability in the morphology of a given type of neuron affected its responses, taking into account the passive cable properties of the neuron's geometry as these properties influence the outcome of local voltage changes. The photo uncaging of glutamate at precise locations on peripheral dendrites provided these voltage events.

Unexpectedly, the study demonstrated that the complicated branching of GM neurons did not lead to the segregation of signals depending on their distance from a somatic recording site. In fact, the neurons' responses resembled those of a single compartment, with fairly uniform membrane potential. In general, such electrotonic compactness would be expected in neurons with few long processes, quite unlike GM neurons. It seems surprising that the intricacies of their different morphologies simply don't affect their function.

The shape of a particular type of neuron arises during the animal's development and is generally expected to be apt for that neuron's physiological functions. But the canonical notions of the optimization of cable properties and the energy efficiency of minimal wiring routes do not seem to apply to these neurons; rather, a "space-filling" mechanism seems to have prevailed. Space-filling neurons develop with numerous and complex neurites that are presumed to maximize the number of processes within a specific volume rather than growing fewer process more directly out to appropriate targets. In the case of stomatogastric ganglion neurons, axons reaching the motor nerves are few compared with the complex tangle of neurites in the neuropil; the stomatogastric neuropil is a striking exemplar of space-filling (see figure 3.3 and plate 5).

In the report, "When Complex Neuronal Structures May Not Matter," the discussion points out that this is "a case in which the solution to the morphology to physiology transform is many-to-one." The various morphologies of the GM neurons studied are good enough but are not optimized for efficiency in "wiring costs." In contrast, many other types of neurons, such as the much-studied CA1 pyramidal neuron of the mammalian hippocampus, differentiate synaptic site locations and compensate for them.

Marder was intrigued, saying, "That is a really interesting piece of work. Computing in analog on a slow timescale doesn't seem to require precise spatial localization because the cell is approximating its potential so it is always compensating. The challenge of getting roughly the right numbers and kinds of ion channels is maintained, but not to micron precision in placing specific channels. We always sort of knew this, but it was in bits and pieces; it's good to have a more direct illustration of it. It's so different from the way people think about vertebrate neurons!"

Around this time and for many months, an awkwardly patchy paper lay on Marder's desk waiting for her guidance. But Marder couldn't think of a story that would hold it together. The report contained data from confocal stacks of numerous dye-filled neurons that had been traced out by several cohorts of the lab's students, with some unfinished quantitative and modeling work intended to analyze the neuronal structures. Evidently, a lot of time and effort had been invested in producing a wealth of information, but somehow the intellectual capital that would bring it all together was missing. Its two principal authors had left the lab and were absorbed in their new jobs elsewhere.

When Otopalik was finishing her uncaging work and writing it up, Marder recognized that her student had developed a real intellectual vision.

"Adriane," said Marder, "make this into a paper."

"And then," Marder relates admiringly, "she tore it to pieces. She did more analysis. She wrote more tools, got the students to retrace images, and remade figures. It was only after doing her first paper that Adriane could understand the significance of this work and what the paper about it should say."

Fortunately, the work had been done on four types of neurons in the same species, *Cancer borealis*, so Otopalik's report, published as "Sloppy Morphological Tuning in Identified Neurons of the Crustacean Stomatogastric Ganglion," could be paired with her earlier paper. The comparison of gastric mill (GM), lateral pyloric (LP), pyloric (PY), and pyloric dilator (PD) neuronal morphologies extended her interanimal comparison of GMs. The quantitative analysis showed that, as in Otopalik's work on the GM neuron, morphology did not appear to be constrained by wiring efficiencies. Again, space-filling development seemed the most likely explanation. And again, there was considerable variability between animals. Interestingly, the study found no reliable morphological characteristics that would distinguish any of the four specific neuron types. There were only two statistically significant dimensions that were specific to two of the cell types: longer total neurite length with higher numbers of branch points in the case of GM neurons, and more symmetry in the ramifications of PD neurons. The evidence supports one of Marder's hypotheses about variability: that properties observed to be variable may just not matter much, and other factors may be more important for maintaining function.

Marder was pleased with the use of new techniques, adding, "Adriane has become an impressive quantitative and innovative deep thinker. A strong scientist. And now that Dave Schulz has the transcriptome, we'll be able to match her methods up with antibody localization to the receptors because we'll be able to find the structures."

In the last few years, new genetic techniques have brought the use of molecular biology to previously recalcitrant organisms like crustaceans. When the time seemed right—and the costs had shrunk—Marder bought the transcriptomes of *Cancer borealis* and *Homarus americanus*. With two species, Marder hoped to be able to look at the same identified cells in both, analyzing the difference in expression levels of similar proteins in them. Comparisons between at least two species may also help in working out which molecules are the most conserved. Now annotated by Schulz, the transcriptomes open up seductive possibilities for Marder. "All of a sudden we can do things in a different way!"

Schulz's work when he was in her lab was a notable feat at the time. He was able to quantify mRNA in identified single cells although for only a small number of channels and cell types. "If we could have done fifty genes on twelve cell types, we would have," Marder says, "but I'm not sure it would have changed our observations at that moment in time. The cell-specific correlations are something we only now start to understand because of Tim's recent work. So we would probably have just seen more correlations but been more confused. We haven't lost anything by waiting five years, because nobody else is further ahead and it would have cost a fortune."

This research has been expanded to a much bigger enterprise. In his own lab, Schulz has been making a detailed comparison of expression levels in identified single neurons, obtaining quantitative PCR measurements on many more receptors and ion channels, with RNA sequencing in the same cells. The earlier PCR work measured the genes for ion channels the lab already knew were there, by reference to the known biophysics, or to work in other invertebrate labs. Now, with the transcriptomes, Schulz has been surveying more broadly and finding things no one knew were there, including numerous receptors and ion channels that have not yet been studied. Marder is jubilant: "We can get the full cast of characters now, all the receptors and ion channels, and then really see what those full sets of correlations look like. I think that's going to be a much more fruitful way to see what kinds of linked systems, correlating systems and compensating systems are there. If you only have a partial cast of characters of anything, it's always a problem."

There are echoes here of Marder's determination to find the full list of peptide neuromodulators, without which she has always thought that deep understanding of the complete phenomenon of neuromodulation operating on the stomatogastric ganglion, an almost completely characterized circuit, would escape her. Today, getting a more complete data set on RNA expression is the first step in deeper studies of compensation. At single-cell molecular level, Marder now feels able to do anything she needs to, and when she has her full cast of characters, the questions she will be able to ask are clear in her mind.

The answer to one important question, Marder suspects, may already be in the data, awaiting analysis, and she is impatient to find out. In Schulz's first studies, although the ratios of expression levels in the different cell types assayed were different, all the cells had approximately the same types of ion channels. Now, able to look at the whole transcriptome, Marder wonders whether this will hold true or whether some types of ion channels will be present in some types of cells but not in others.

Working with Schulz, she wants to acclimate animals to high and low temperatures and compare full sets of channel expression in their neurons. That work will first focus on constant temperature, on getting the molecular and expression data analyzed. Then, going back to perturbation experiments, they will be able to study the kinds of molecular changes that underlie compensation, with a much higher level of resolution in terms of the number of receptors and channels, the better to see the full pattern.

Marder is also excited about the opportunity to study receptor genes to find out, cell type by cell type, exactly what receptors are present in each. Then she wants to get precise localization of modulators acting on receptors, using photo uncaging of peptides and amines. I asked her how she approached the experimental designs when she doesn't know what's there or what she is missing or not testing for, and she replied, "I don't worry about that. If one is going to do anything in experimental science, you start where you can. You have no choice if you want to go forward; you have to be courageous enough to go forward in the most sensible way at the time. Otherwise we might as well do nothing."

* * *

For years, Marder has argued that homeostasis predicted compensation; homeostasis led to the multiple solutions for maintaining function, observed experimentally and computationally, and these were possible only if mechanisms for compensation existed in both single neurons and networks. Now, compensation has become the big story, and Marder is nowhere near the end of her quest to understand its mechanisms. The principal question occupying her thoughts is whether compensation is optimized and, if so, in what way. It's a Russian doll of a question, concealing several other layers.

Are some individuals in a species robust in the face of some perturbations but less robust to others? Is there a hard set of constraints that comes into play? Or are some individuals just more robust across the board? As Marder asks, "Are some animals just healthier than others, that is to say, more proficient at being crabs than others? Are the people who live to be a hundred just physiologically more robust human beings than the people who die at seventy?" She adds that since all human beings contend with multiple environmental and internal and external stresses, it is important to ask what it would mean to be better optimized for a range of constraints. Can the compensations necessary to confront one kind of perturbation coexist with the compensations for another—do they synergize or do they interfere? Is there a hierarchy of compensations: is one type of compensation more important than others—is it essential for a certain species to be good at responding to a particular perturbation? Does evolution select for one or for many such properties? Or do individual animals within a species operate a compromise between compensations? Faced with this complex set of interlocking questions, Marder remains sanguine: "In my mind the problems are very well defined. Whether it will be possible to get true experimental clarity on them—that's a separate issue."

Framing those questions in both theoretical and experimental terms so as to separate out the underlying factors of compensation has become her major concern for the next few years. Models and experiments have to be designed to investigate whether those animals that are most robust to extreme temperature are also robust to pH changes and then to look at how those same animals endure changes in oxygen tension. In the ocean, those conditions are not unrelated, as oxygen tension and pH change with temperature, but in the lab they can be uncoupled. Then, after finding the phenomena, all the molecular processes that might be involved will have to be identified and described.

Marder concedes that there are difficulties: "The biggest is how to design experiments so you're not fooled by lack of stationarity. Ideally, and as usual, we want to do multiple things on the same preparation, but you have to face lack of stationarity issues. It's a challenge but solvable." When a researcher observes a neuron in a preparation, it is a snapshot, at that moment in time, of its possible range of properties. "It's drifting around. Then you perturb it and knock it to some other state. So the question is, when you perturb it for the second or third time, will you get the same response or not. If not, it's hard to compare two different perturbations. It might take a little fiddling, but one always has that kind of problem."

I once asked Marder an ill-defined question about the inherent characteristics of stomatogastric neurons: of the thirty neurons, how many might be inherently excitable? She answered, "That's one of those questions! It depends. If you do the same experiment over and over again, if you isolate the LP cell, sometimes it'll be silent, sometimes it'll be firing tonically. Sometimes you'll hyperpolarize it and it'll go into plateau, sometimes it won't. We've never been able to get a handle on the modulatory history and the underlying state of each neuron. There's a lot of unknowable history, and unknowable damage; you can't read it out."

* * *

An additional set of questions concerns failures of compensation, which can be catastrophic for an animal, resulting in disease, pathology, or death. Initially, it was difficult to say whether succumbing to acute perturbations, such as high temperature, was caused by a failure of homeostasis, of regulation, or of compensation.

Recent work, both theoretical and experimental, in Marder's lab implies that some perturbations can result in pathological states even when the homeostatic mechanisms are working correctly. A homeostatic system may have mechanisms that normally provide stability and can respond quite well to many but not all perturbations. Some particular perturbation may take the system into a state where the normal homeostatic mechanism is working correctly but the system as a whole doesn't get back to its original behavior.

This is shown in computational work, in particular in O'Leary's models. The regulation rules are chosen in such a way that compensation for the sorts of perturbation an organism is likely to undergo acts to maintain some property, such as the rhythm of a pyloric circuit. However, if a perturbation that the organism is never likely to encounter arises, a deletion of a certain gene for instance, then the same compensatory mechanism may sometimes work, sometimes not. Sometimes it may make matters worse, the compensatory mechanism itself being at fault. It is not a failure, though, of the regulatory system, which is still intact, but is now operating on a different system that this regulatory system has not evolved to cope with. Therefore, its regulation of, say, ion channels is no longer appropriate, and the neuron might no longer function. That could be the basis for seizures or some loss of function in the nervous system.

Understanding compensation and its failures will, Marder thinks, shed light on a question she has pondered since her doctoral research: how can the boundary between health and disease be identified? Given that there are evident differences in the properties of what are called "normal" brains, at what point do differences in those underlying structures and their operations cross the line to abnormality, affecting, say, the control of excitability or seizure mechanisms? For this reason, her studies of compensation are of interest to researchers in the fields of epilepsy, neuropathic pain, and injuries to the spinal cord that can cause upregulation in reflex pathways.

Marder does not suggest that the organization of the mammalian spinal cord is going to be similar to the simpler pattern generating circuits of the stomatogastric ganglion. Even in different species of invertebrates, although some may display similar-looking motor patterns, the circuits that produce them have turned out to be structured differently. Moreover, the equivalent pattern-generating circuits in mammals contain numerous interneurons; the alternation that results from reciprocal inhibition in invertebrates may be produced by quite different mechanisms. It is a big leap from a primitive cord like the lamprey's to that of a quadruped and another leap to the spinal cord of bipeds; walking upright on two feet involves all sorts of issues of balance and coordination, nothing like moving in water.

Over the past few years, Marder has thought more, and more deeply, about large, mammalian circuits and the relevance of her concepts to them. She has always faced resistance to her ideas on the grounds that they are not generally applicable, and not translatable to mammalian neuroscience in particular, because of the peculiarities of the stomatogastric ganglion preparation. Marder sums it up with some resignation: "I get told it's a circuit that's controlling a stomach, therefore the rules are different from circuits for cognition. And I just say 'Huh?'"

The prevailing belief has been that invertebrate neurons are different: different in morphology, different in their summation strategies, different in requiring so many neuropeptide modulators, whereas vertebrate preparations must use other mechanisms because they have so many more neurons. But as Marder ruefully points out, "Every biological phenomenon we've studied, somebody would tell me that large vertebrate circuits didn't need to do that because they have lots of neurons. So vertebrate neurons don't do bursts—they can get by with spiking because there are so many of them. Or they don't have co-transmitters, or they don't have plateaus, they don't have graded transmission ... plus this lingering prejudice that if you want to understand large networks you really only have to understand synaptic connectivity. Every one of these arguments has turned out to be wrong."

Marder readily acknowledges that there are indeed many questions about large networks that cannot be answered by results from invertebrate neuroscience, although she complains that these questions are still poorly defined. "I keep waiting for someone to tell me something about the way very large ensembles work in language that's different from the three- and the five-cell circuit language."

Perhaps the most fundamental of the unanswered questions is this: compared with small circuits, what are the special features of large networks associated with their size, and how do they arise? Is there the equivalent of a phase transition somewhere between small and large networks?

"There are bound to be some things you can do with a large network that you can't do with a small network," Marder says, "but I don't think anyone has a very good grasp on why you need a network of this particular size to do that particular task. Why, in the vertebrate brain, do you have five hundred neurons doing this, and a million and a half doing that, and ten million doing the other? What is it about those computations or those functions that requires that size or that kind of computational complexity in a circuit?"

Nevertheless, whatever the specific properties of large networks turn out to be, Marder expects the understanding of how network behavior depends on specific cellular mechanisms to be relevant to circuits of all sizes. In fact, as researchers go from single-unit recordings with single electrodes to hundreds of thousands of neurons with tetrodes to record their coordinated and correlated activity, attention to the detail at cellular level may turn out to matter a lot.

Marder is excited and enthusiastic about the technological advances in neuroscience, from the new understanding of the genetic and molecular basis of cell organization to optical recording from large brain ensembles, and, of course, the tremendous developments in computer technology that process and allow interpretation of astonishing amounts of data. Her enthusiasm is tempered by a concern that some of the lessons she and her colleagues have learned may not be heeded in the rush to exploit the new techniques. She hopes that researchers will recognize the confound of parallel pathways, saying, "In the field of small circuit cracking, which some of us grew up in, we worked hard to get functional connectivity maps. Circuits are highly interconnected and drastically reconfigured by neuromodulation. One of the things we learned was that there were often lots of parallel pathways and multiple ways that cells communicated. So it was often very difficult to figure out how information was moving. Now, all of a sudden, say you're working in a mouse model and you can use an optogenetic tool, or a genetic deletion, you can do a single manipulation, get a result, and assume that you have the answer. Whereas you may have one of five answers. Or a whole series of half answers that are interpreted as if they're full answers. And likewise you can delete a cell type or a specific pathway and see no effect, but that doesn't mean that they aren't important for some part of that circuit."

In contrast to the thirty named neurons of the stomatogastric ganglion, the millions of neurons to be classified by cell type in large vertebrate systems evidently present a challenge. Typically they are very small, entangled together, not necessarily morphologically similar within cell types, and their position relative to other neurons is variable.

In the stomatogastric ganglion, with its large neurons, cell-type identification depends primarily on the anatomical projection patterns. Then there are characteristic electrophysiological phenotypes (although these can change), and soon there will be characteristic molecular phenotypes (also subject to variation). But even in this small system, it would be rash to rely on either electrophysiology or molecular properties alone.

It could be said that a new era, a molecular era, has opened to the stomatogastric ganglion. No sane researcher would ever have set out to make transgenic crustaceans. But new tools for editing DNA sequences, such as transcription activator-like effector nucleases (TALENS) and clustered regularly interspaced short palindromic repeats (CRISPR), together with engineered viruses, have made it possible to skip the process of making a mouse or drosophila line, achieving the same goals with one shot. So it is now possible to study fundamental molecular biology questions in crustaceans, and because of the size of its cells and full complement of identified cell types, the stomatogastric ganglion has gained in attraction. It is a cold preparation and relatively slow, which is helpful in the lab. Because the system is known in more detail than most and is well quantified, it is promising for studying spatial as well as temporal gene regulation. There are definitely simpler neurons to be found, but the geometrical complexity of stomatogastric ganglion neurons does not necessarily entail biochemical complexity. Above all, it is a really robust preparation. It is possible to plan a manipulation taking two weeks, such as monitoring the movements of different RNAs in neurites, which would be difficult in other systems.

* * *

For forty years, Marder's lab has been collecting data from thirty neurons simultaneously recorded with several electrodes. "We have zillions of miles of data, and that's not enough. Just having the anatomy isn't enough. You have to know how to think about things, which means you have to frame questions. Any amount of data in the absence of clearly posed questions won't tell you anything. Except that sometimes the questions are revealed when you just look at the data."

This is an archetypal Marder principle and key to what I have learned about her distinctive approach. Always alert to the implications of anomalous data, she observes as much evidence as she can lay her hands on and always has done so. Over forty years, that includes how the animals live, their natural behavior, the properties of the preparations in the lab, the results of experimental interventions on them, the influence of their genes, and the possibilities and constraints that computational methods point to. Then she thinks. The habit of not rushing to the bench was formed in San Diego and has been vindicated ever since. Schooled in the manner of our older generations to appreciate history and value the roots of her science, she is convinced of the importance of thought. She is forever impressed by the early discoveries in science that were revealed by penetrating thought backed up by what we now see as rudimentary equipment and techniques.

Some of Marder's thinking skills, with knowledge and experience, have been built up over forty years, but essentially she started out with a remarkable talent for reasoning in long lines of thought. She may have stayed faithful to a narrow path, but her eyes have always been open to the landscape beyond that path. She reads widely, and her curiosity has never waned. She is always eager for the opportunity to discuss her science, and other people's or other fields.

So how does she think? Like a detective, but a detective without bias. The data provide the evidence, and Marder never strays far from the plot as revealed by that evidence. She analyzes and muses on her data with a mind perhaps not untrammelled by preconceptions, but she is able to set them aside. The data always hold the key. That is how she was able to recognize the evidence pointing to variability, to take just one example of her many insights, only some of which are included in this book.

Marder has always used the small stomatogastric nervous system to answer big questions, saying, "Obviously at some level I knew where I was going from the beginning. But on another level it was a real act of faith to think that there was an experimental path from studying transmitters in the ganglion controlling stomach movements in a lobster to understanding schizophrenia—and we don't yet understand that."

The lobster is still teaching lessons; its stomatogastric ganglion is still relevant and rich in possibilities, not least the possibilities of mining its history of research, the data compiled over forty years. Marder calmly agrees: "It is comforting to know there remain fundamental and general unanswered questions that can be profitably addressed with a beautiful ganglion of only thirty neurons."

Acknowledgments

My gratitude to Eve Marder is impossible to express adequately, and believe me, my thesaurus is well thumbed. Obviously, this book could not have been written without her generous cooperation and far too much of her time. Thank you.

I am very grateful for the enthusiasm and patience of the scientists I interviewed and regret that the book couldn't include many more of the stories and insights that were shared. This list is in alphabetical order:

Larry Abbott, PhD, Professor of Neuroscience and Co-Director of the Center for Theoretical Neuroscience, Columbia University, New York.

Carol Barnes, PhD, Professor of Psychology, University of Arizona, Tucson.

Dirk Bucher, PhD, Associate Professor of Biological Sciences, New Jersey Institute of Technology, Newark.

Ron Calabrese, PhD, Professor of Neuroscience, Emory University, Atlanta.

Patsy Dickinson, PhD, Professor of Neuroscience, Bowdoin College, Brunswick, Maine.

Judith Eisen, PhD, Professor of Neuroscience, University of Oregon, Eugene.

Marie Goeritz, PhD, Research Fellow, Marine Science, University of Auckland, New Zealand.

Jorge Golowasch, PhD, Professor of Biological Sciences and Mathematics, New Jersey Institute of Technology, Newark.

Gabrielle Gutierrez, PhD, Research Associate, Department of Applied Mathematics, University of Washington, Seattle.

Scott Hooper, PhD, Professor, Department of Biological Sciences, Ohio University, Athens.

Lingjun Li, PhD, Professor of Pharmaceutical Sciences and Chemistry, University of Wisconsin, Madison.

Brian Mulloney, PhD, Professor of Neuroscience, University of California, Davis.

Farzan Nadim, PhD, Professor of Biological Sciences and Mathematics, New Jersey Institute of Technology, Newark.

Mike Nusbaum, PhD, Professor of Neuroscience, University of Pennsylvania, Philadelphia.

Timothy O'Leary, PhD, Lecturer in Information Engineering and Medical Neuroscience, University of Cambridge.

Maria Pellegrini, PhD, Executive Director of Programs, W. M. Keck Foundation, Los Angeles.

Astrid Prinz, PhD, Associate Professor of Biology, Emory University, Atlanta.

David Schulz, PhD, Principal Investigator, Schulz Lab, University of Missouri, Columbia.

Sonal Shruti, PhD, Post-Doctoral Researcher, Institut National de la Santé et de la Recherche Médicale, Université de Lille, France.

Karen Sigvardt, PhD, Adjunct Professor of Neurology, University of California, Davis.

Frances Skinner, PhD, Professor of Computational Neuroscience, University of Toronto.

Gina Turrigiano, PhD, Professor of Biology, Brandeis University, Waltham.

I have enjoyed every minute spent in Professor Arthur Wingfield's company and greatly appreciate his kindness and wit.

My husband, Jon Foulds, a sharp-eyed grammarian and critic, although not a scientist, has improved the text and put up with a lot of angst. My sister, Liza Cody, a fiction writer, improved my narrative style and constantly cheered me on. I set much store by comments from my brother, Dr. Michael Nassim, the family's polymath. Michael Z. Lewin has made many valuable suggestions. Thank you all for your generosity and affection.

Robert Prior, Executive Editor for Life Sciences and Neuroscience at the MIT Press, has tolerantly passed over the missed deadlines and always stayed as positive about the book as he was when I first approached him out of the blue.

Glossary

acetylcholine   Usually referred to as ACh, acetylcholine is a neurotransmitter and neuromodulator in both vertebrates and invertebrates.

action potential   An essential mechanism in neuronal signaling, the action potential is a short-lived voltage change in membrane potential that is propagated along a neuron's axon once the membrane potential has passed a threshold. Canonical values for this change in membrane potential are from about -70 mV to $+30$ mV.

agonist   An agonist is a substance that can bind to and activate the same receptors as a substance under study; an antagonist binds to the same receptor as that substance but blocks the receptor's response. Pharmacologists have found a library of substitute molecules that can block or activate specific receptors. These substances are used in laboratories if they are more stable or can be more easily purified or synthesized than the naturally occurring molecules.

annotation (see transcriptome annotation)

cybernetics   A classical Greek term meaning "governance," reintroduced by Norman Wiener in his 1948 book of which the title says it all: "Cybernetics: Control and Communication in the Animal and Machine." Cybernetics is applicable to systems arising in many disciplines such as engineering, biology, artificial intelligence, and so on. It is principally expressed in mathematical models.

decapod crustaceans   Lobsters and crabs, crayfish and shrimps, are decapod crustaceans; they have shells and five leg-like appendages on each side. They all have stomatogastric ganglia, although obviously of different sizes and shapes.

depolarization   Neurons, like other cells, have a negative charge inside the cell relative to the extracellular environment. Depolarization occurs when the negative charge becomes less negative or transiently positive, as, for example, during an action potential.

electrical synapses   Also known as a gap junction, electrical synapses pass ions directly between two neurons and are found at closely apposed axons and dendrites. They are faster than chemical synapses, and most of them are bidirectional. Electrical coupling is widely used in both invertebrate and vertebrate nervous systems.

electrotonic   In a neuron, this refers to local changes in membrane potential that attenuate as they spread and do not contribute to the generation of an action potential.

gap junction (see electrical synapses)

graded potential   Graded potentials arise from the summation of the individual actions of ligand-gated ion channel proteins and decrease over time and distance. Depending on the stimulus, graded potentials can be depolarizing or hyperpolarizing. The amplitude of a graded potential is proportional to the strength of the stimulus, unlike the all-or-nothing action potential.

half-center oscillator   Two reciprocally coupled nonbursting neurons that produce a rhythmic output constitute a half-center oscillator. For example, if the two neurons have inhibitory synapses together, one neuron's firing can inhibit the other, then release it from inhibition, at which point the second neuron fires, inhibiting the first.

hemolymph   The fluid circulated by an arthropod's heart; it contains nutrients, hormones, cells (hemocytes), and oxygen, which is carried by hemocyanin, a molecule that includes copper.

homeostasis   Generally, the term refers to the maintenance of stable physiological conditions in an animal's body—its *milieu intérieur*, as the French physiologist Claude Bernard (1813–1878) called it. It can also refer to the active regulation of any of the animal's physiological systems. The term "homeostasis" was introduced by the American physiologist Walter Cannon (1871–1945).

hormone   A hormone is a chemical substance secreted by a gland or cell. It may act by diffusion into local tissue or be carried by the bloodstream, producing more general effects. Hormones exert their influence by binding to receptors in the cell membrane.

interneuron   An interneuron has connections with other neurons but none with effectors (such as muscle). Interneurons tend to be found in ganglia or nuclei. They are small and have multiple dendrites and axons, but these do not form long projections.

ions   Ions are atoms or molecules in solution that have lost or gained one or more electrons. As electrons are negatively charged, the loss of an electron produces a positively charged ion; examples in this book are sodium ions (Na^+) potassium (K^+), and the doubly charged ion of calcium (Ca^{2+}). The chloride ion gains an electron and therefore has a single negative charge (Cl^-).

iontophoresis   Iontophoresis is used to apply a small amount of a substance such as a neurotransmitter to a tiny area of a cell, such as a site close to a synapse or a neuromuscular junction. The substance is contained in a micropipette, and a contact wire introduces an electrical charge that drives the substance out of the pipette.

The micropipette can be combined with a recording electrode. Iontophoresis can be contrasted with "bath" applications of a pharmacological substance to a preparation such as a stomatogastric ganglion in a dish.

linear   Relationships that can be represented as a straight line on a graph are linear. In general parlance, linear processes are understood to mean those that follow a stepwise sequence.

machine code or machine language   The essential level of language for the instructions that computers can obey, machine code is in number form (binary or hexadecimal). It is not readable by sane humans, but it is fast because it does not require the computer to translate from word based instructions as it goes along. Computer programming generally uses higher level languages such as Python, MATLAB, Pascal, or Java, which can be written and read as quasi languages, usually based on English.

MALDI   Matrix-Assisted Laser Desorption/Ionization is a mass spectroscopy technique for identifying the individual molecules in complex mixtures.

monosynaptic connection   A monosynaptic connection between two neurons is direct: no interneuron is connected between them.

mRNA   Messenger ribonucleic acid is one of the three principal types of RNA; the other two are transfer RNA (tRNA) and ribosomal RNA (rRNA). Transcribed from DNA in the nucleus, mRNA carries the code sequences for amino acids to ribosomes, where tRNA provides those component amino acids to form specific proteins.

mRNA expression   An animal's DNA is the same in all its nucleated cells, but different cell types use different selections of genes from their DNA to make the proteins they need. These genes are translated into RNA, and the abundance of different mRNAs thus "expressed" can be measured to provide a snapshot of the cell's current state of activity.

neuroethology   A portmanteau word formed from neurobiology and ethology, neuroethology is the discipline of studying and comparing the operations of nervous systems and the natural behavior they give rise to in different animals.

neuropil   Any area in a nervous system that consists of a dense concentration of unmyelinated neuronal processes (axons and dendrites) allowing multiple synapses between relatively small numbers of neurons is called a neuropil.

parameter space   Mathematical models of neurons or networks use ranges of values for parameters such as conductances. For example, a model neuron may be represented by a set of different conductances and each of these conductances has a range of values; the set of all combinations of those values is called the parameter set. These values can be plotted on a two- or three-dimensional graph that illustrates the parameter space and differences between regions in that space. The term is used in many other contexts.

PCR   The polymerase chain reaction (PCR) is a technique used to make millions of copies (known as amplification) of a DNA sequence for study.

pharmacology   Pharmacology investigates how chemical substances exert their effect on the physiology of living organisms.

phenotype   From personality to hair color, phenotype refers to the characteristics constituting an individual organism, produced by the genotype and the environment.

plateau firing   A membrane depolarization above the firing threshold can persist when a synaptic input or brief depolarization causes lasting inward currents so that a neuron continues to fire without further synaptic input.

postsynaptic potential (PSP)   A postsynaptic potential is a transient change in the polarization of a neuron's membrane potential caused by synaptic input from another neuron. It can be excitatory or inhibitory. The PSPs of a neuron are summated. If the result depolarizes the membrane potential over its threshold, then an action potential may be produced.

power spectrum   A power spectrum reveals the distribution of power in a waveform at its different frequencies. Familiar examples include power spectrum analyses of the sounds of musical instruments, revealing their distinctive timbres.

Q_{10}   In biological processes (or chemical reactions), Q_{10} is the rate change factor for each increase in temperature of 10°C. It is an indication of how temperature dependent a process is: a Q_{10} of 1 indicates temperature invariance; higher values indicate increasing rate change with rising temperature. Many physiological processes have a Q_{10} of around 2, which means that their rates double with a 10°C rise in temperature.

radio-immuno assay   This method can sensitively measure minute quantities of molecules such as hormones or neurotransmitters in a liquid preparation. It uses competition for the corresponding antibody between the substance to be assayed and radio-labeled isotopes of it.

RNAseq   RNA sequencing is a technique used to detect the presence and measure the quantity of RNA in a tissue sample. Its results are specific to that particular moment in time, so it can be used to monitor changes in gene expression, for example, during different experimental interventions.

statistical power   The formal definition states that statistical power is the probability of rejecting the null hypothesis when it is, in fact, false. The user-friendly definition is the probability that an experiment has distinguished a real effect rather than chance effects. The power of a statistical test is influenced by the size of the effect to be observed, the number of observations, and the level of statistical significance chosen for the research.

stomatogastric ganglion neurons   The numbers of each neuron type commonly found in the ganglion are given in parentheses if more than one.

Neurons of the pyloric circuit:

AB, anterior burster
PD (2), pyloric dilator
LP, lateral pyloric
PY (6–9), pyloric
VD, ventricular dilator
IC, inferior cardiac

Neurons generally classified as participating in the gastric circuit:

GM (3–7), gastric mill
LPG (2), lateral posterior gastric
LG, lateral gastric
MG, median gastric
DG, dorsal gastric
AM, anterior median
Int1, interneuron 1

Other neurons:

CD2, cardiac sac dilator, part of the cardiac sac circuit, found in the stomatogastric ganglion of the spiny lobster and probably of other species.
AGR, anterior gastric receptor, a sensory neuron just outside the stomatogastric ganglion but often counted with its neurons.

stomatogastric nervous system   The stomatogastric nervous system consists of paired bilaterally symmetrical knots of nerves, the commissural ganglia, a single esophageal ganglion, and the stomatogastric ganglion. The stomatogastric nerve from the esophageal ganglion is the principal input to the stomatogastric ganglion.

striated muscle   Striated muscle is sometimes called striped muscle or, in vertebrates, skeletal muscle. This type of muscle is similar in arthropods and vertebrates. More than 700 million years ago, the common ancestor of arthropods and vertebrates must have had striated muscle. Invertebrates other than arthropods have only smooth muscle (sometimes called "visceral" in vertebrates).

transcriptome   Named in an analogy to the genome, which is an organism's complete set of DNA, its genes, the transcriptome is its set of RNA molecules, which transcribe the DNA. The term can be used to refer to the complete complement of RNA in an organism's cells or to mRNA in a single cell.

transcriptome annotation   The analysis of the functions and biological processes associated with each of the proteins encoded by an organism's RNA.

References

Eve Marder's full publication list is available on line; only papers mentioned in the book are listed here. Publications by other scientists to which reference is made are included. For each chapter, references are in first author alphabetical order.

Chapter 1 The Lone Reader

Sir John Eccles. (1965) The synapse. *Scientific American, 212*(1), 56–66.

Chapter 2 First Findings

Bruner, J., & Kennedy, D. (1970). Habituation: Occurrence at a neuromuscular junction. *Science, 169,* 92–94.

Hanley, M. R., & Cottrell, G. A. (1974). Acetylcholine activity in an identified 5-hydroxytryptamine-containing neurone. *Journal of Pharmacy and Pharmacology, 26,* 980.

Hanley, M. R., Cottrell, G. A., Emson, P. C., & Fonnum, F. (1974). Enzymatic synthesis of acetylcholine by a serotonin-containing neurone. *Nature: New Biology, 251,* 631–633.

Kehoe, J. (1972). Three acetylcholine receptors in *Aplysia* neurones. *Journal of Physiology, 225*(1), 115–146.

Marder, E. (1974). Acetylcholine as an excitatory neuromuscular transmitter in the stomatogastric system of the lobster. *Nature, 251,* 730–731.

Otsuka, M., Kravitz, E. A., & Potter, D. D. (1967). Physiological and chemical architecture of a lobster ganglion with particular reference to gamma-aminobutyrate and glutamate. *Journal of Neurophysiology, 30*(4), 725–752.

Parker, G. H. (1948) *Animal Colour Changes and their Neurohumours: a Survey of Investigations 1910–1943.* Cambridge: Cambridge University Press. (New edition 2012).

Remler, M., Selverston, A., & Kennedy, D. (1968). Lateral Giant Fibers of Crayfish: Location of somata by dye injection. *Science, 162*(3850), 281–283.

Chapter 3 Lobster Lore

Bucher, D., Johnson, C. D., & Marder, E. (2007). Neuronal morphology and neuropil structure in the stomatogastric ganglion of the lobster, *Homarus americanus. Journal of Comparative Neurology, 501*, 185–205.

Krogh, A. (1929). The Progress of physiology. *American Journal of Physiology, 90*(2), 243–251.

Maynard, D. M. (1955). Activity in a crustacean ganglion. II. Pattern and interaction in burst formation. *Biological Bulletin, 109*(3), 420–436.

Selverston, A. I., & Moulins, M. (Eds.). (1986). *The Crustacean Stomatogastric System, A Model for the Study of Central Nervous Systems.* Springer Verlag.

Chapter 4 Marking Time

Greengard, P. (2003). Nobel prizewinner's lecture. In H. Jörnvall (Ed.), *Nobel Lectures, Physiology or Medicine 1996–2000.* Singapore: World Scientific.

Kehoe, J. S., & Marder, E. (1976). Identification and effects of neural transmitters of invertebrates. *Annual Review of Pharmacology and Toxicology, 16*, 245–268.

Marder, E. (1976). Cholinergic motor neurones in the stomatogastric system of the lobster. *Journal of Physiology, 257*, 63–86.

Marder, E., & Paupardin-Tritsch, D. (1978). The pharmacological properties of some crustacean neuronal acetylcholine, gamma-aminobutyric acid and l-glutamate responses. *Journal of Physiology, 280*, 213–236.

Chapter 5 A Lab of One's Own

Beltz, B., Eisen, J. S., Flamm, R., Harris-Warrick, R., Hooper, S. L., & Marder, E. (1984). Serotonergic innervation and modulation of the stomatogastric ganglion of three decapod crustaceans (*Panulirus interruptus, Cancer irroratus, and Homarus Americanus*). *Journal of Experimental Biology, 109*, 35–54.

Eisen, J. S., & Marder, E. (1982). Mechanisms underlying pattern generation in lobster stomatogastric ganglion as determined by selective inactivation of identified neurons. *Journal of Neurophysiology, 48*, 1392–1415.

Eisen, J. S., & Marder, E. (1984). A mechanism for the production of phase shifts in a pattern generator. *Journal of Neurophysiology, 51*, 1375–1393.

Hooper, S. L., & Marder, E. (1987). Modulation of a central pattern generator by two neuropeptides, proctolin and FMRFamide. *Brain Research, 305,* 186–191.

Marder, E. (1984). Mechanisms underlying neurotransmitter modulation of a neuronal circuit. *Trends in Neurosciences, 7,* 48–53.

Marder, E., Calabrese, R. L., Nusbaum, M. P., & Trimmer, B. (1987). Distribution and partial characterization of FMRFamide-like peptides in the stomatogastric nervous systems of the rock crab, *Cancer borealis,* and the spiny lobster, *Panulirus interruptus. Journal of Comparative Neurophysiology, 259,* 150–163.

Marder, E., & Eisen, J. S. (1984). Transmitter identification of pyloric neurons: Electrically coupled neurons use different neurotransmitters. *Journal of Neurophysiology, 51,* 1345–1361.

Marder, E., & Eisen, J. S. (1984). Electrically coupled pacemaker neurons respond differently to the same physiological inputs and neurotransmitters. *Journal of Neurophysiology, 51,* 1362–1374.

Marder, E., & Paupardin-Tritsch, D. (1980). The pharmacological profile of the acetylcholine response of a crustacean muscle. *Journal of Experimental Biology, 88,* 147–159.

Selverston, A. (1976). A model system for the study of rhythmic behavior. In J. C. Fentress (Ed.), *Simpler Networks and Behavior* (pp. 82–98). Sunderland, MA: Sinauer.

Chapter 6 The Multifunctional Network

Dickinson, P., Mecsas, P., & Marder, E. (1990). Neuropeptide fusion of two motor-pattern generator circuits. *Nature, 344,* 155–158.

Getting, P. A. (1989). Emerging principles governing the operation of neural networks. *Annual Review of Neuroscience, 12,* 185–204.

Getting, P. A., & Dekin, M. S. (1985). Tritonia swimming: A model system for integration within rhythmic motor systems. In A. I. Selverston (Ed.), *Model Neural Networks and Behavior* (pp. 3–20). New York: Plenum.

Heinzel, H.-G., Weimann, J. M., & Marder, E. (1993). The behavioral repertoire of the gastric mill in the crab, *Cancer pagurus*: An in situ endoscopic and electrophysiological examination. *Journal of Neuroscience, 13*(4), 1793–1803.

Hooper, S. L., & Marder, E. (1987). Modulation of the lobster pyloric rhythm by the peptide proctolin. *Journal of Neuroscience, 7*(7), 2097–2112.

Marder, E., & Hooper, S. (1985). Neurotransmitter modulation of the stomatogastric ganglion of decapod crustaceans. In A. I. Selverston (Ed.), *Model Neural Networks and Behavior* (pp. 319–337). New York: Plenum.

Weimann, J. M., Meyrand, P., & Marder, E. (1991). Neurons that form multiple pattern generators: Identification and multiple activity patterns of gastric/pyloric neurons in the crab stomatogastric system. *Journal of Neurophysiology*, *6*, 111–122.

Chapter 7 Asking the Right Question

Buchholtz, F., Golowasch, J., Epstein, I. R., & Marder, E. (1992). Mathematical model of an identified stomatogastric ganglion neuron. *Journal of Neurophysiology*, *67*(2), 332–340.

Golowasch, J., & Marder, E. (1992). Ionic currents of the lateral pyloric neuron of the stomatogastric ganglion of the crab. *Journal of Neurophysiology*, *67*(2), 318–331.

Kepler, T. B., Marder, E., & Abbott, L. F. (1990). The effect of electrical coupling on the frequency of model neuronal oscillators. *Science*, *248*(4951), 83–85.

LeMasson, G., Marder, E., & Abbott, L. F. (1993). Activity-dependent regulation of conductances in model neurons. *Science*, *259*, 1915–1917.

Chapter 8 Tuning to Target

Goldman, M. S., Golowasch, J., Marder, E., & Abbott, L. F. (2001). Global structure, robustness, and modulation of neuronal models. *Journal of Neuroscience*, *21*, 5229–5238.

Golowasch, J., Abbott, L. F., & Marder, E. (1999). Activity-dependent regulation of ionic currents in an identified neuron of the stomatogastric ganglion of the crab *Cancer borealis*. *Journal of Neuroscience*, *19*(RC33), 1–5.

Golowasch, J., Casey, M., Abbott, L. F., & Marder, E. (1999). Network stability from activity-dependent regulation of neuronal conductances. *Neural Computation*, *11*, 1079–1096.

Golowasch, J., Goldman, M. S., Abbott, L. F., & Marder, E. (2002). Failure of averaging in the construction of a conductance-based neuron model. *Journal of Neurophysiology*, *87*, 1129–1131.

Liu, Z., Golowasch, J., Marder, E., & Abbott, L. F. (1997). A model neuron with activity-dependent conductances regulated by multiple calcium sensors. *Journal of Neuroscience*, *18*(7), 2309–2320.

Siegel, M., Marder, E., & Abbott, L. F. (1994). Activity-dependent current distributions in model neurons. *Proceedings of the National Academy of Sciences of the United States of America*, *91*, 11308–11312.

Turrigiano, G., Abbott, L. F., & Marder, E. (1994). Activity-dependent Changes in the Intrinsic Properties of Cultured Neurons. *Science*, *264*, 974–977.

Turrigiano, G., LeMasson, G., & Marder, E. (1995). Selective regulation of current densities underlies spontaneous changes in the activity of cultured neurons. *Journal of Neuroscience, 15*(5), 3640–3652.

Chapter 9 Good Enough

Bucher, D., Prinz, A. A., & Marder, E. (2005). Animal-to-animal variability in motor pattern production in adults and during growth. *Journal of Neuroscience, 25,* 1611–1619.

Goaillard, J.-M., Taylor, A. L., Schulz, D., & Marder, E. (2009). Functional consequences of animal-to-animal variation in circuit parameters. *Nature Neuroscience, 12,* 1424–1430.

Grashow, R., Brookings, T., & Marder, E. (2009). Reliable neuromodulation from circuits with variable underlying structure. *Proceedings of the National Academy of Sciences of the United States of America, 106,* 11742–11746.

Grashow, R., Brookings, T., & Marder, E. (2010). Compensation for variable intrinsic neuronal excitability by circuit-synaptic interactions. *Journal of Neuroscience, 30*(27), 9145–9156.

Prinz, A. A., Billimoria, C. P., & Marder, E. (2003). An alternative to hand-tuning conductance-based models: Construction and analysis of data bases of model neurons. *Journal of Neurophysiology, 90,* 3998–4015.

Prinz, A. A., Bucher, D., & Marder, E. (2004). Similar network activity from disparate circuit parameters. *Nature Neuroscience, 7,* 1345–1352.

Schulz, D. J., Goaillard, J.-M., & Marder, E. (2006). Variable channel expression in identified single and electrically coupled neurons in different animals. *Nature Neuroscience, 9,* 356–362.

Schulz, D. J., Goaillard, J.-M., & Marder, E. (2007). Quantitative expression profiling of identified neurons reveals cell-specific constraints on highly variable levels of gene expression. *Proceedings of the National Academy of Sciences of the United States of America, 104,* 13187–13191.

Chapter 10 In the Big Picture

Gutierrez, G. J., O'Leary, T., & Marder, E. (2013). Multiple mechanisms switch an electrically coupled, synaptically inhibited neuron between competing rhythmic oscillators. *Neuron, 77,* 845–858.

Hamood, A.W., Haddad, S.A., Otopalik, A.G., Rosenbaum, P. and Marder, E. (2015). quantitative reevaluation of the effects of short- and long-term removal of descending modulatory inputs on the pyloric rhythm of the crab. *eNeuro, 2*(1), 0058–14.

Hamood, A. W., & Marder, E. (2015). Consequences of acute and long-term removal of neuromodulatory input on the episodic gastric rhythm of the crab *Cancer borealis*. *Journal of Neurophysiology*, *114*, 1677–1692.

Marder, E., O'Leary, T., & Shruti, S. (2014). Neuromodulation of circuits with variable parameters: Single neurons and small circuits reveal principles of state-dependent and robust neuromodulation. *Annual Review of Neuroscience*, *37*, 329–346.

O'Leary, T., Sutton, A. C., & Marder, E. (2015). Computational models in the age of large datasets. *Current Opinion in Neurobiology*, *32*, 87–94.

O'Leary, T., Williams, A. H., Caplan, J. S., & Marder, E. (2013). Correlations in ion channel expression emerge from homeostatic regulation mechanisms. *Proceedings of the National Academy of Sciences of the United States of America*, *110*, E2645–E2654.

O'Leary, T., Williams, A. H., Franci, A., & Marder, E. (2014). Cell types, network homeostasis and pathological compensation from a biologically plausible ion channel expression model. *Neuron*, *82*, 809–821.

Otopalik, A. G., Goeritz, M. L., Sutton, A. C., Brookings, T., Guerini, C., & Marder, E. (2017). Sloppy morphological tuning in identified neurons of the crustacean stomatogastric ganglion. *eLife*, *6*, e22352.

Otopalik, A. G., Sutton, A. C., Banghart, M., & Marder, E. (2017). When complex neuronal structures may not matter. *eLife*, *6*, e23508.

Tang, L. S., Goeritz, M. L., Caplan, J. S., Taylor, A. L., Fisek, M., & Marder, E. (2010). Precise temperature compensation of phase in a rhythmic motor pattern. *PLoS Biology*, *8*(8).

Tang, L. S., Taylor, A. L., Rinberg, A., & Marder, E. (2012). Robustness of a rhythmic circuit to short- and long-term temperature changes. *Journal of Neuroscience*, *32*(29), 10075–10085.

Index